Easy-to-Make
WOODEN SUNDIALS

Instructions and Plans for Five Projects

With Suggestions for Designing
Your Own Pocket Sundial

Milton Stoneman

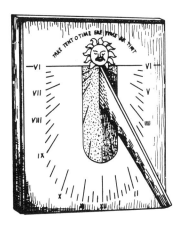

Dover Publications, Inc.
New York

To Edna, for listening

Copyright © 1982 by Milton Stoneman.
All rights reserved under Pan American and International Copyright Conventions.

Published in Canada by General Publishing Company, Ltd., 30 Lesmill Road, Don Mills, Toronto, Ontario.
Published in the United Kingdom by Constable and Company, Ltd.

Easy-to-Make Wooden Sundials is a new work, first published by Dover Publications, Inc., in 1982.

Book design by Carol Belanger Grafton
Figure illustrations by Janette Aiello

International Standard Book Number: 0-486-24141-6
Library of Congress Catalog Card Number: 81-67086

Manufactured in the United States of America
Dover Publications, Inc.
180 Varick Street
New York, N.Y. 10014

CONTENTS

LORE & LEGEND	page 6
MATERIALS & CRAFT	7
FIRST PROJECT: HORIZONTAL SUNDIAL	9
SECOND PROJECT: VERTICAL SUNDIAL	15
THIRD PROJECT: FOLDING EQUATORIAL SUNDIAL	19
FOURTH PROJECT: DIPTYCH SUNDIAL	23
FIFTH PROJECT: BOWSTRING EQUATORIAL SUNDIAL	27
DESIGN YOUR OWN POCKET SUNDIAL	29
CORRECTING YOUR SUNDIAL	31
LIST OF SELECTED CITIES	34
MOTTOES	37
ADDITIONAL READING	38

Templates for the five projects (Plates I–VI) follow page 20.

LORE & LEGEND

When it comes to sundials, there is nothing new under the sun. The science of sundials (called *gnomonics*, from the Greek word *gnōmōn* or interpreter; the gnomon is the part of the dial that casts a shadow) is the same today as when it was first conceived millennia ago. The only difference between ancient and modern dials is not in their principles but that, with our modern tools and mathematics, dials are easier to make today.

Chances are that if you ask anyone what a sundial is, you'll get a description, complete with illustrative hand motions, of the common garden variety of dial: a flat plate ringed with roman numerals and a "thing" (the gnomon) sticking up in the middle that casts a shadow and tells the time — if the sun is shining. That's about right. The ordinary dial is a simple two-part device: a flat surface marked with the hours and a gnomon slanted at the angle of the dial's latitude (the distance in degrees from the equator) and pointed toward the North Star. The only moving part is the earth. You may also be familiar with vertical dials mounted on walls facing south (in the Northern Hemisphere). Whether horizontal or vertical, once set in place the dial becomes a time-telling instrument for here and now — here, because it was necessarily designed to be accurate at only one location, and now, because the shadow tells you the local sun time.

Less familiar than horizontal and vertical dials are any number of imaginative gadgets that interpret the celestial mechanics of the dynamic earth-sun relationship. You can see a Roman hemicycle in New York's Central Park; it is mounted on a semicircular stone bench whose curved back casts changing seasonal shadow lines on the calibrated pavement. A bejeweled equatorial dial adorns the entrance to the Mexico City airport. There is a different type of equatorial dial in Denver's Cranmer Park; another variation on the beach of English Bay, Vancouver, British Columbia; and a giant modern horizontal dial in Sun City, Arizona. A famous vertical sundial is placed on the wall of Queens College, Cambridge, England; one of its features is a chart of figures for each segment of the moon's monthly period, making it a moondial that can tell time in the moonlight. A hole in a cathedral dome in Florence, Italy, lets in a noon sunbeam to shine on the floor, thereby forming a sundial. Sightseers delight in the dial near the Rock of Gibraltar. An ancient dial in Jerusalem tells time by using an old system with hours that vary in length.

Why so many dials? Because they served a purpose; even today they exercise a certain charm on users of battery-powered digital watches. Prehistoric man created the first solar timepiece by poking a stick into the ground, dividing daylight into forenoon and afternoon; this gave him ample warning to return to his cave before darkness set in and beasts came out. Dials became more sophisticated when the need arose for more accurate timepieces. A tenth-century Saxon "scratch" dial indicated no more than the "tide" or time of daily Masses. The European Renaissance entailed increased social and commercial activity, world exploration, and advances in art and education; hence sundials were created that made finer divisions of the day and that told the time not only for "here and now" but also for other places. Dialists competed for royal commissions; talented craftsmen worked in wood, silver, ceramics, and stone, putting dials on items as diverse as pistol butts and finger rings; every gentleman's education included lessons in gnomonics and practical sundial construction.

And then there were clocks. Sundials receded rapidly into the realm of ornamentation, where they are found today, of interest to the traveler, scholar, antiquarian, romantic, craftsman, and collector. Accuracy and convenience have been gained, but something has been lost as well; for sundials, old-fashioned though they are, are direct reminders of what we mean by time: the measurement of human activity in terms of the positions of heavenly bodies.

MATERIALS & CRAFT

Let's make a sundial. First let it be understood that this is not the usual how-to treatise. You don't have to be a master craftsman to follow the plans in this book. Making these projects is like working with a ready-mix: all you have to do is pick out the parts you need, many of which may be purchased already cut to size, and assemble them according to the plan you choose and your latitude. I have gone into extra detail on the first and simplest of the projects, the horizontal sundial; what you learn on that project will guide you as you move on through the book.

This is not to say that sundial theory and construction can't get pretty complicated. It can; libraries are full of books on how to plot the dial's hour lines, sometimes requiring trigonometric calculations for angles, lines, and curves. How interesting, informative, and helpful these books are depends on the reader's understanding and craftsmanship. *Easy-to-Make Wooden Sundials* takes care of the intimidating complexity for you. Rather than requiring you to calculate the precise measurements for your location (remember that a dial must be designed for a particular latitude), I have developed measurements for four different latitudes (25°, 32°, 40°, and 45°) that correspond roughly to the latitudes of Miami, New Orleans, New York, and Montreal, making this book useful for all of the United States and a good part of the rest of North America. The templates have measurements for the four latitudes. All you have to do to find your latitude is to look at Figure 24 and the List of Selected Cities. (Use the city on the list that is closest to where you live; the approximate latitude will be accurate enough for our projects.)

For tools and materials you can do without an electronic calculator, parallel rules, and complicated tables. You may use a protractor for checking your angles as you work. For the most part the only tools you will need for planning the dial are a metal-edged ruler, an inexpensive draftsman's compass (the kind used in grammar school art classes is fine), pushpins, a pencil, a ball-point pen, and carbon and plain paper. You will need some clear pine shelving stock, a hobby saw (hand or electric), a wood-burning tool or "electric pen," such as the kind Boy Scouts use, for marking the hour lines and inscriptions, a matte knife (for example, an X-ACTO knife), and some coarse and fine sandpaper. Should you choose to make a metal gnomon (see the chapter on vertical dials), a hacksaw blade and metal file will be necessary. You can weatherproof the dial with outdoor spar varnish. If the dial is to be kept indoors as a conversation piece, treat its base to a felt underside.

Although my aim was to develop a how-to book that anybody could work with, the resourcefulness and skill of the craftsman may suggest other, more sophisticated, materials and tools. A talented woodworker may choose to work with harder wood, develop a new shape for the base, and put a special emphasis on staining and finishing. Similarly, someone who works with metal, stone, or plastics may be able to bring to the projects an innovative aesthetic sensibility. Regardless of the medium used, the plans and fundamental methods in this book will serve. The principle of sundials never varies, but sundial construction is as various as the people who make them.

Dial decoration is another area where your own imagination and skill can play a creative role. The shape of a dial's base, particularly for the vertical dial, may be square, round, or like a heart; it may be very long and narrow or short and wide, depending on the space it must fit into; it may be shaped like a shield, an apple, a guitar, or any form of your design. Regardless of the shape of the base, the dial will be accurate provided that the hour angles, which radiate from the dial's center and extend to the edge of the base you have designed, are as the plans prescribe.

A word about using the templates. The templates for the five projects are gathered at the center of the book, following page 20, and are numbered Plates I–VI. Turn to the middle page of the book (Plate VI will thus be spread open before you) and pry open the binding staples. Do not break the staples or remove them entirely. When the staples are open, you will be able to remove the six plates from the book. Do this carefully so that the paper does not tear. With all the plates set apart from the rest of the book, bend the staples back into place, restoring the book to its original shape (except for the removed templates).

It is very important that you cut out the templates carefully—the accuracy of your sundial depends on it. Use scissors or a matte knife, depending on which instrument you feel most comfortable with. It is a good idea to do some practice cutting before you begin to work on the templates. Don't cut out any templates, however, until you read *all* the instructions for the project you are working on.

When you have the templates for a particular project cut out, I recommend that for most of the projects you transfer the patterns to the wood by using the *carbon-paper transfer method*. This consists of placing a piece of carbon paper, carbon-side down, on the wood and putting the pattern on top of the carbon paper. You can keep the pattern and carbon paper in place with pushpins or tape. Be very careful not to let the pattern slip because that will introduce errors into the dial's face. Use a ballpoint pen or sharp pencil for tracing the lines of the pattern. For curves, an appropriately set and positioned draftsmen's compass will provide you with the clearest tracings. When you finish tracing, remove the pattern (storing it for future use) and carbon paper. There is another transfer method, the *pushpin transfer method*, that you will find useful. It is described in detail for the fifth project. Regardless of the method you employ, when you finish the transfer you will see the sundial pattern on the wood.

A helpful hint about storing the templates: label each template carefully. If you don't do this, you may find when you return to the templates months later that you can't tell, for example, whether the template for a dial face belongs to the horizontal or vertical project. When you cut the templates out, you will find that you are cutting away a great deal of information. Transfer all this information to the proper templates.

FIRST PROJECT

HORIZONTAL SUNDIAL

Materials

- 11″ × 11″ piece of wood or a wooden plaque for the dial face
- 5/16″-thick piece of wood for the gnomon
- Pencil
- Metal scribe
- Metal-edged ruler
- Matte knife
- Saw (electric preferred)
- Electric pen (wood-burning tool)
- Fine sandpaper
- Medium wood file
- Draftsman's compass
- Magnetic compass
- Stain
- Shellac
- Carbon paper
- Ball-point pen
- Tape
- Carpenter's level
- Pushpins

The templates (Parts A through C) for this project appear on Plates I and II. As mentioned previously, there are templates for four different latitudes. This is where the pleasure and interest of the mechanics of dialing come in. Inasmuch as the dial will be made for a specific location, select the templates that are best suited for your latitude (see the List of Selected Cities). The dial face (Part A) is the same for all latitudes. If you live near Chicago (41°50′), use the plans for 40°, that is Part A (dial face), Part B₂ (hour lines), and Part C₂ (gnomon). If you live near Miami, use the 25° plans (B₄ and C₄). If you're near Seattle or Montreal, the 45° plans (B₁ and C₁) are for you. You can also find the latitude of your town in an atlas.

Note that the plans have two kinds of lines: dotted lines are for where you will cut the wood, solid lines are for where you will mark the wood with the electric pen or the matte knife. If you want to make a six-sided (sexagonal) dial, select a piece of wood without knots, streaks, or blemishes. Ponderosa pine cuts easily. Hardwoods require more care and effort. Cut the wood to an 11″ × 11″ square. Don't make the square into a sexagon until *after* the various steps are completed and the gnomon is fitted. If you want to have a circular dial plate, you can go to any hobby or crafts store and buy a wooden plaque. These plaques are already sanded, nicely shaped, and beveled. They are ready to use as is.

Finding the Center of the Dial Plate

If you are using an 11″ × 11″ square board, all you have to do to find the center is draw the two diagonals. The center of the plate is where the diagonals intersect. This point is not to be confused with the *dial center*, which is the midpoint of the 6:00 line. The 6:00 line will be discussed later.

If you are using a store-bought round plaque for the dial plate, you can locate the center in any of several ways, including any number of methods you may remember from your high-school geometry class. I will explain one very simple method here. Take a sheet of paper that is larger than the plaque. Lay the paper on the plaque. Holding the paper in place with one hand, run your finger firmly over the plaque's circumference. The crease in the paper forms a circular pattern the size and shape of the plaque. Cut out the pattern and lay it over the plaque. The pattern and the plaque should align perfectly. Fold the paper *exactly* in half and place it once again on the wood. Use the fold as a guide to draw a light pencil line on the wood. The line is a diameter of the circular board; its midpoint is the center of the plaque. To find the midpoint, turn the paper a few degrees in either direction and draw another diameter. The intersection of the two diameters is the center of the board. You can verify this with your draftsman's compass, or you can use a ruler to confirm that the intersections are indeed midpoints.

Transferring the Dial Face *(Part A)* to the wood

Cut out the template for the dial face (Part A). It

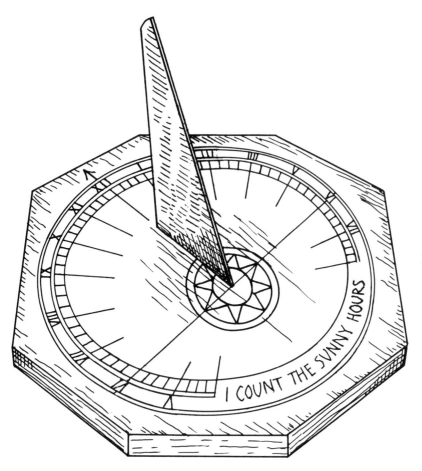

Figure 1. Six-sided horizontal sundial with metal gnomon.

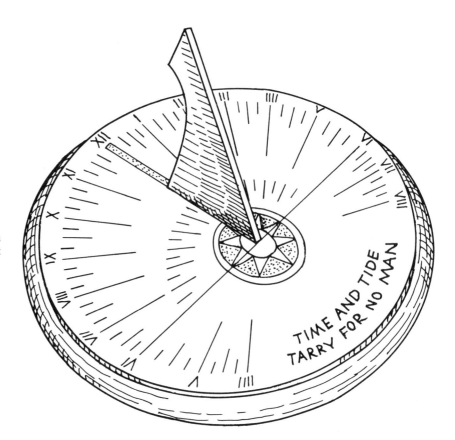

Figure 2. Horizontal sundial with wooden gnomon and store-bought round plaque.

10 *Easy-to-Make Wooden Sundials*

should be noted here that all the measurements of Part A were calculated with a wooden gnomon $5/16''$ thick in mind. This measurement is noted on the plan; it is the distance between Points *LX* and *RX*. (Point *X* is the midpoint of the 6:00 line. *RX* is $5/32''$ to the right of *X*; *LX* is $5/32''$ to the left.) If you use a thinner piece of wood for the gnomon, or if you do not use wood at all but metal (perhaps fashioned from a wall bracket) or opaque plastic, you will have to change the distance between *LX* and *RX* accordingly.

Use a fresh sheet of carbon paper to transfer the plan to the wood. The carbon paper goes carbon-side down on the wood. Put a pushpin through the carbon paper and into the center of the dial plate. Use transparent tape to fix the carbon paper securely to the wood. Then take the Part A template and plate it over the carbon paper. (You will have to remove the first pushpin.) It is very important that the center of the dial plate on the template aligns exactly with the actual center of the wood plate. Tape or pin the template snugly over the carbon paper and to the wood. See that neither the carbon paper nor the template moves or slides from its fixed position.

Carefully trace every line and mark on the template with a ball-point pen. Maintain the distinction between broken and solid lines—otherwise you may saw off an essential part of the dial. When you are finished tracing, the dial face will be clearly outlined on the wood. Save the template for future horizontal sundial projects. If you handle it carefully, the carbon paper can be used once or twice again.

Transferring the Hour Lines *(Part B)*

Select and cut out the appropriate hour-line template (B_1, B_2, B_3, or B_4) for your latitude. Note that Points *RX* and *LX* appear on both Part A and Part B; these points are $5/16''$ apart—precisely the width of the wooden gnomon. The two vertical lines that rise from *RX* and *LX* are the *double 12:00 lines*. When the dial is completed and the gnomon is in place, the sun will cast a sharp shadow on the left side of the gnomon in the morning and on the right side in the afternoon. If you use a thin metal gnomon, the double 12:00 lines will not be required —a single 12:00 line will do.

It is very, very important that you lay the hour lines exactly in position on the wood. Be sure the 6:00 lines on the template and the wood align; check the double 12:00 lines also. When you are certain the template is in place, push three pins through the paper and into the wood at Points *LX*, *RX*, and *Y*. Note that no carbon paper is used for the transfer; rather you will use a ruler to extend the template's lines beyond the edge of the template and onto the wood face. The *LX* and *RX* pins will guide your ruler as well as hold the template in position. Use little bits of tape to hold down the outer edges of the template, but be careful not to put the tape where hour lines are to be ruled.

Now you are ready to mark the hour lines. The morning hours will be on the left, the afternoon hours on the right. Start with the morning hours, beginning at 11:00 and moving down toward the 6:00 line that is already on the board. Five o'clock on the morning side will be an extension of 5:00 on the afternoon side, and 7:00 on the afternoon side will be an extension of 7:00 on the morning side. Place your ruler snugly against the pin at *LX* and along the 11:00 line. Take a sharp pencil and gently but firmly draw a line on the wood from the outermost circle (Circle A) to the template. Now move your ruler to the 10:00 line and draw a line from the appropriate point on Circle *A* to the template. Draw in the lines for 9:00, 8:00, and 7:00 also.

The half-hour lines (for 6:30, 7:30, 8:30, and so forth) should be drawn shorter than the hour lines. Without moving the template, look again at the carbon tracings of the dial face. Instead of starting at Circle *A* and drawing a line to the template, start at Circle *B* and draw your lines only as far as Circle *C*. (As you can see from the carbon tracings, *B* and *C* are not actually full circles, but I call them circles for the sake of simplicity.) Once again use gentle but firm pressure to pencil in the lines, always making sure your ruler fits snugly against Point *LX*.

After you have completed all the hour and half-hour lines for the morning side of the dial, draw in the lines for the afternoon side. The procedure is the same as for the morning side except that your ruler is held snugly against the pin at Point *RX*. Don't forget to extend the 5:00 P.M. and 7:00 A.M. lines to draw the lines for 5:00 A.M. and 7:00 P.M. respectively. Now you have completed your work with the Part B template. With all the hour and half-hour lines in place, you can mark off the quarter-hour lines by dividing each half-hour by eye. Note that the lines between 11:00 and 12:00 will be closer together than those between 6:00 and 7:00.

Burning in the Lines

I should mention at this point that although I have been using arabic numerals for the hours (6:00, 7:00, etc.) in our discussion, I prefer to use roman numerals (VI, VII) when I actually burn in the hours on the dial. The reason for this is that all of us are more familiar with arabic numerals; but when it comes to the dial itself, I like the antique charm of the roman numerals. Roman numerals are also easier than arabic to burn in with an electric pen. The four-stroke roman IIII, rather than IV, is customarily used on sundials.

All the lines, circles, and roman numerals on the dial face are burned in with an electric pen. If you have never used this tool before, carefully read the instructions that come with it. It's best to practice with the tool on a piece of scrap wood. With a little experience you will learn what is the most comfortable way for you to hold the electric pen and how much pressure is necessary to mark the wood in a straight line. Try a few curves and roman numerals. It's important that you *control* the pressure so that the grain of the wood does not carry the tool away from the desired path.

When you have practiced enough, burn in the markings on the dial. Work slowly and carefully, and *stay on the line*. The reward of your care is the neatness and accuracy of your finished dial.

Constructing the Gnomon *(Part C)*

The template for the gnomon (Part C), like the template for the hour lines, has plans for four different latitudes. Choose the pattern that is for the correct latitude for you. Cut out the template and use the carbon-paper transfer method to apply the pattern to the wood.

The wood used here is 5/16″-thick clear pine. Check to see that this thickness is exactly the same as the distance between the double 12:00 lines on the dial face. With your saw, cut the gnomon out of the wood. On the template you will see that one end of the gnomon drawing is a shaded stub with broken outlines. The stub is the part of the gnomon that will be inserted into and below the dial face and will not be visible when the sundial is completed. For now, simply cut along the dotted lines of the stub and along the solid lines for the rest of the gnomon.

Now for the slot in the dial face. Guide your metal scribe against the metal-edged ruler to score the lines for the dial slot (where the gnomon stub will be inserted). If you're not sure which lines to use, look again at the template for the dial face, or the carbon tracing of that template on the board. The shaded rectangular area around the dial plate center is the slot where the gnomon stub will be inserted. Carefully chip away at the wood with a matte knife until the slot is the right length, width, and depth to accommodate the inserted gnomon stub. Use your wood file to "square up" the corners of the gnomon stub until you get a snug, perfect fit in the slot. The gnomon will be properly glued in place later.

The fitted gnomon should stand firmly and precisely between the double 12:00 lines. The lower end of the gnomon's slanting edge should meet the dial face exactly at the 6:00 line. The broad side of the gnomon — that is, *not* the side that is 5/16″ wide — should be at a right angle to the dial face (see Figure 3).

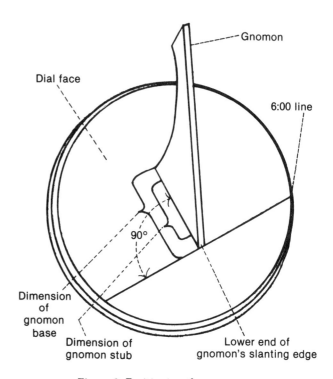

Figure 3. Positioning the gnomon.

To check the alignment of your gnomon, use the gnomon template to construct a stiff cardboard duplicate of the wooden gnomon but *without* the stub. Figure 4 shows what a miniature cardboard replica of a 45° gnomon would look like. Note that Points X′ and T′ on the cardboard replica correspond to X and T on the wooden gnomon. Place the piece of cardboard against the inserted wooden gnomon. The cardboard and wood should match up exactly (though it's only the angle of the straight top of the gnomon that determines the dial's accuracy). There are many ways to check to see that the broad side of the gnomon does indeed form a right angle with the dial base; the easiest is to take a piece of cardboard that you know is square and place it in the angle between the gnomon and dial face.

Finishing the Horizontal Sundial

Don't glue the gnomon into place yet. If you used an 11″ × 11″ board for the dial face, now is the time to make it into the sexagon of the template for the dial face. Round off the corners with a wood

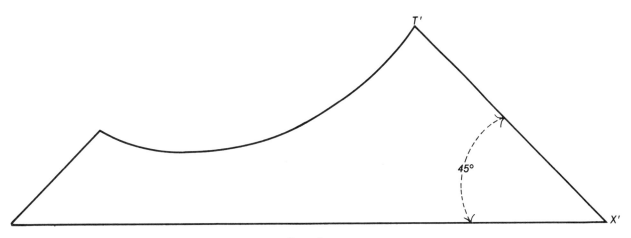

Figure 4. Cardboard testing gnomon.

file. Sand all surfaces until they are smooth and even to the touch. If the wood burning was done properly, all pencil lines and finger marks can be sanded away, leaving the brown, burned lines. If you used the store-bought plaque instead of the square board, you probably will have little or no finishing to do to the dial face.

Do not round off the edge of the gnomon's slanting side. The sharp, straight edge casts the shadow that falls on the dial face's hour lines.

This is also the time to burn in an appropriate motto for your dial. Many suggested mottoes are appended to this book. Pencil in your motto on the dial, leaving ample and uniform space between letters. Then burn in the letters. The care you bestow on the motto will distinguish your sundial from all others.

If you are imaginative, you can add your own touches to the completed dial. The template I have provided calls for a sunburst design at the base of the gnomon, but you can do additional wood burning and carving to heighten your dial's look of antiquity and charm.

Glue the gnomon into place. After the glue dries, stain the wood *lightly*. (A dark stain will obscure the shadow.) Then apply several coats of shellac, allowing drying time between coats, to give the dial a rich finish as well as protection from the weather.

Setting Your Sundial

Choose a place for your dial that is fully exposed to the sun. Check out the location for a few sunny days to make sure that it receives the morning, noon, and afternoon sun without the interference of a tall tree or building. Watch especially in the early morning and late afternoon, when the sun's low, slanting rays can be cut off by a garage or a neighbor's roofed patio.

When you have selected the best possible place for the dial, the dial is ready to be set. It is very important that you place the dial on a platform or pedestal that is perfectly level. Use a carpenter's level or bubble to check the platform from front to back, side to side, and even from corner to corner. Now you may place the dial on the platform.

As mentioned earlier, the gnomon must always point north. Take a small magnetic compass and place it on the dial base just behind the gnomon. Turn the compass (not the sundial base) so that the "north" mark points directly in line with the gnomon. Unless you have had the remarkably good fortune to place your dial facing north by chance, you must now set the compass and, with it, the dial. Use both hands to rotate the dial plate gently, being careful not to let the compass slide or turn. When the compass needle points to the "north" marking, stop turning the dial and rest it on the platform. Now refer again to the List of Selected Cities and note the last column for the magnetic variation for your city. (An explanation of magnetic variation appears in the chapter on correcting your sundial.) The dial is now set.

If using a magnetic compass seems like an intrusion of the modern world into the venerable art of "dyalling," you could orient your dial as the ancients did, without using a compass. Take a flat board and draw several concentric circles on it. The diameters should be approximately 3", 3½", and 4". Lightly drive a long finishing nail (the smaller the head, the better) into the center of the circles. The nail should be perpendicular to the board.

Place the board on the level surface where you intend to mount your sundial. About 1½ or 2 hours before noon, note the exact moment when the tip of the nail's shadow just touches one of the circles. Mark this point with a pencil dot. As the time gets closer to noon and the sun mounts higher in the sky, the nail's shadow will shorten and move closer to the center of the circle. After noon the shadow will begin to lengthen on the other side of the nail.

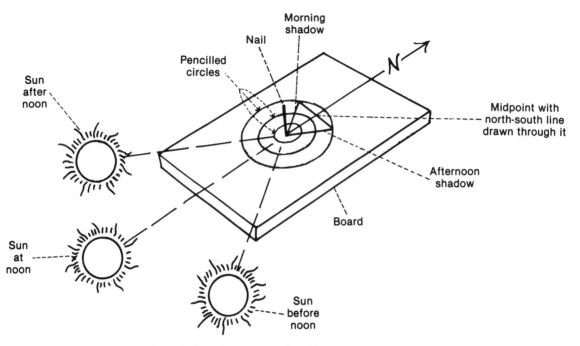

Figure 5. Locating true north without a compass.

Watch for the moment when the tip of the shadow touches the same circle on the other side. Mark the point of intersection with your pencil. It is very important that you do not shift the board from its original position.

Now draw a straight line between the two marked points. Find the midpoint of the line. Then draw a line from the center of the circle through the line's midpoint and beyond (see Figure 5). This line points true north; you may use it to orient your sundial in place of a magnetic compass.

At this point you could install the dial permanently by bolting it in position to the platform, making sure that the dial plate remains level and the gnomon continues to point north. Before you do that, however, I recommend that you read the chapter on correcting your sundial. There are several ways to improve the accuracy of your dial; for example, if you don't adjust for what is called the *equation of time* and your longitude, you will probably find that your watch and your sundial don't agree. That's because, in a special sense, your sundial is more accurate! Time is the measure of the relation between the earth and the sun, and that is what your dial tells, *sun time*. The time you hear on the radio, what you set your watch to, is *standard time*, a figure arrived at by averaging the various sun times over a broad area (called a *time zone*). Standard time is useful for coordinating all sorts of activities. The school bell rings at 9:00, standard time; the pupils would never arrive at school together if each of them lived by his own sundial. A football field that runs east-west has a different sun time at either end. An explanation of how to go from sun time to standard time is provided in the section on correcting your dial—though standard time, strictly speaking, is not a correction but an "incorrection."

SECOND PROJECT

VERTICAL SUNDIAL

This vertical dial is often called a *direct south vertical sundial* because it is designed to be mounted vertically on a wall that faces directly south. There also are dials called *vertical decliners*—dials that face generally south but "decline" a little to the east or west. Vertical decliners require special calculations for their construction. We will not deal with these dials here.

Materials

The materials for the vertical and horizontal dials are similar, but note these important exceptions. For the vertical dial you will need a 7" × 10" board. Either you can cut this board carefully from clear lumber, or you can buy a wooden plaque from a crafts store. Instead of using a wooden gnomon, for the vertical dial I used a metal shelving bracket. These brackets are readily available in any hardware or home-furnishing store. An 8" bracket will do.

Procedure

The templates (Parts A through C) for this sundial appear on Plates II and III.

As you work on the vertical dial, keep in mind the instructions given with the horizontal dial for transferring the templates and cutting and assembling the pieces. Note that the vertical and horizontal dials differ in four important ways.

First, the vertical dial must be mounted perfectly vertically. You can check this with a simple weighted plumb line.

Second, the wall must face directly south. Use a small magnetic compass to check this out. If the wall does not face directly south, place a piece of wood of the correct thickness behind the dial plate, thereby angling the dial properly.

Third, the angle of the vertical dial's gnomon is the complement of the latitude of the dial's location—that is, the angle is 90° minus the latitude. As with all the plans in this book, of course, no calculations are necessary on your part. The gnomon, like all the other parts, has been precalculated and the proper dimensions have been incorporated into the template.

Finally, the vertical sundial must be read counterclockwise. (The horizontal dial was read clockwise.)

You must transfer the dial face template (Part A) to the wood with carbon paper. In order to align the template properly, it is necessary to measure in from the side of the panel to determine the midpoint. Then lightly pencil in a vertical line dividing the plate in half. This will be the dial center line and the 12:00 line. Measure 2¼" down along the vertical line from the top edge of the plate; mark that point, then draw a horizontal line through it, from the left edge to the right, exactly at a right angle to the vertical line. This horizontal line is the 6:00 line.

You are now ready to transfer the dial face to the wood. Align the template carefully with the wood. Make sure the two center Points X match up. Check the alignment of the horizontal and vertical lines also. With a ball-point pen trace all the guidelines and the sunburst design. The guidelines will serve to position the roman numerals and hour lines. When you remove the template and carbon paper (storing them for future use), the outline on the board should look just like the template.

Next select the correct template for the hour lines (Part C). Keep in mind the discussion of latitudes in the section on the horizontal sundial. Transfer the template to the wood as you did for Part C of the horizontal dial—with a metal-edged ruler, pencil, and pushpins. Note that there is a single 12:00 line. There were double 12:00 lines for the horizontal dial because you had to accommodate the thickness of a wooden gnomon, but for the vertical dial you are using a metal gnomon of negligible thickness. If you choose to use a wooden gnomon, you will have to make a double 12:00 line equal to the thickness of the gnomon wood, as on the horizontal dial. The

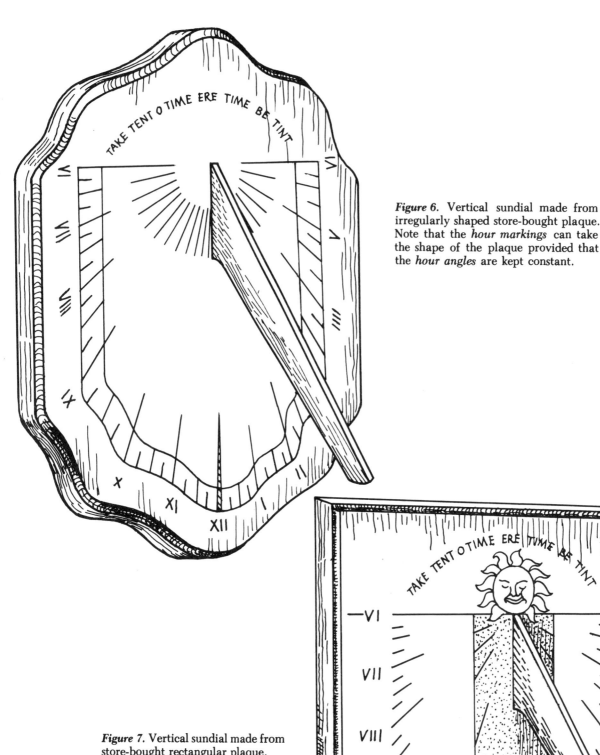

Figure 6. Vertical sundial made from irregularly shaped store-bought plaque. Note that the *hour markings* can take the shape of the plaque provided that the *hour angles* are kept constant.

Figure 7. Vertical sundial made from store-bought rectangular plaque.

16 *Easy-to-Make Wooden Sundials*

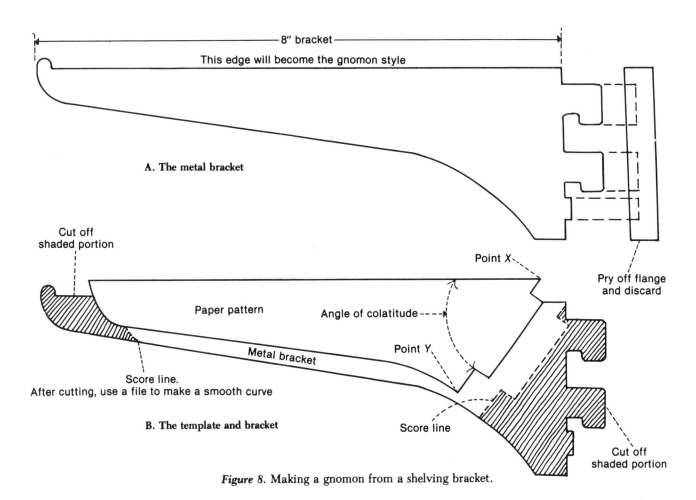

Figure 8. Making a gnomon from a shelving bracket.

morning hour lines, of course, would be drawn on the left side of the vertical double 12:00 lines and the afternoon hour lines would be drawn on the right side of the vertical double 12:00 lines. As always, work slowly and carefully. Slight errors in hour lines and angles become gross when the lines are extended from the central origin.

The gnomon used for the original model of this vertical dial was made from an 8″ metal shelving bracket. These brackets are available in all hardware and home-furnishing stores. There are many advantages to using a store-bought bracket. First, the style (the upper slanting edge of the gnomon) will be perfectly straight. Second, the metal is rigid; thus the shadow cast by the gnomon will be clear and accurate. Third, the metal is easy to work: you can pry off the flange with pliers, and a hacksaw will meet with very little resistance when you cut a straight line. More advanced hobbyists, of course, may wish to work with rigid sheet metal.

Choose the correct gnomon template for your latitude. Transfer the pattern to the metal bracket either by drawing its outlines on the metal, or by tracing it onto a piece of carboard. I prefer the latter method because the cardboard serves as a guide when you score the metal. Before you score the metal, however, it is important that you check the alignment of the cardboard and the metal bracket —style edge must align with style edge. Of course the line between Points X and Y should go at the wide end of the bracket (see Figure 8).

Use your matte knife to score the metal. The shading in Figure 8 indicates what area is unwanted; cut this area off with a hacksaw. Use a metal file to smooth off the edges and to shape a smooth curve at the end of the gnomon. Do not touch the straight shadow-casting edge of the gnomon.

Now for the slot in the dial face. Use your matte knife to slice out the slot neatly. The fit of the gnomon in the slot should be snug so that the broad side of the gnomon remains fixed at a right angle to the dial face. The base of the gnomon (Point X) should just touch the center of the 6:00 line. Point Y on the gnomon should line up with Point Y on the dial face, on the vertical center line. With the gnomon thus in place, the gnomon's angle is equal to the complement of your latitude.

The dial is now ready for testing, decorating, and finishing.

Figure 9. Folding equatorial sundial. View of upper (summer) dial face.

Figure 10. Folding equatorial sundial. View of lower (winter) dial face and bottom tablet with gnomon slot.

18 *Easy-to-Make Wooden Sundials*

THIRD PROJECT

FOLDING EQUATORIAL SUNDIAL

Unlike the other dial projects in this book, which have plans for four different latitudes, the folding equatorial dial can be adjusted to seven latitudes simply by sliding the brass (nonmagnetic) gnomon in or out. This adjustable feature makes the dial portable. The equatorial dial is also relatively easy to plan. It requires no complicated calculations, unlike the horizontal and vertical dials (though the complicated aspects of the horizontal and vertical dials were taken care of prior to the creation of the templates, which incorporate the solutions to the mathematical puzzles).

The equatorial dial consists of two hinged wooden panels. The top panel has two dial faces on it, an upper and a lower; you have templates for both faces (Plate IV). The sun casts a shadow on the upper face from March to September, and on the lower face from September to March. This is so because the sun is said to be *on* the equator at the vernal and autumnal equinoxes (March 21 and September 23 respectively), *above* the equator from March to September, and *below* the equator from September to March. The equatorial dial is designed so that the top panel simulates the earth's equator. Thus when the sun is above the equator, it is also above the top panel (and casts its shadow on the upper face); when the sun is below the equator, it is also below the top panel (and casts its shadow on the lower face).

Figure 11 illustrates the mechanics of the equatorial dial for 40° latitude (the latitude of New York). The dial plans have been designed so that, for your chosen latitude, the top panel will always be parallel to the earth's equator, and the gnomon will

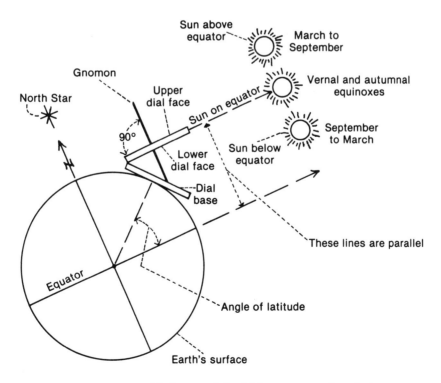

Figure 11. Mechanics of the folding equatorial sundial.

always point exactly north—to the polestar. There is a negligible inaccuracy in this kind of sundial: for purposes of design and construction, it must be assumed that the location of your dial on the earth's surface and the center of the earth 4,000 miles away are one and the same point. How can 4,000 miles be negligible? Because the point of reference is the sun, 93,000,000 miles away, where the light that casts your dial's shadow originates. If you could stand on the sun and look at the earth, the earth would be a tiny speck, if it were visible at all; and the 4,000 miles between the earth's center and surface could not be discerned.

Materials

Besides many of the items required for the previous projects, in order to make the equatorial sundial you will need two identical pieces of clear pine 5" × 8" × ⅜", a brass rod for the gnomon ⅛" in diameter and 8⅝" long, and two brass (non-magnetic) hinges that measure ¾" × 1" when open. You will also need a drill and an inexpensive plastic protractor to check the angles.

Transferring the Dial Faces to the Wood

First make the dial's upper panel. Set one 5" × 8" board before you, with the larger dimension running vertically. Divide the shorter side in half, and draw a line from top to bottom through the horizontal midpoint. Measure up 3" along this line from the bottom; mark the point, then draw a line through it at a right angle to the vertical line. The second, horizontal line should extend to the right and left edges of the board.

Now follow the same procedure for the lower face. Turn the board over and measure and draw your lines. It's important that you designate the same end of the board as the bottom as you did for the upper face. The 3" points on the two faces must align. Later you will drill a hole through these points.

Return to the upper face. Cut out the appropriate template and place it over the wood. Make sure that the vertical center line at the bottom of the template corresponds exactly with the pencilled center line on the wood. Also check the alignment of the horizontal line through the 3" point. With the template in position, place a pushpin through the template's center dot and firmly into the wood. Lift the template carefully to see that the pin is in the wood precisely at the intersection of the vertical and horizontal lines. When everything is in place, tape the template to the wood on the left, right, and bottom (*not* the top). Slip a piece of carbon paper under the template and trace the sunburst design. If properly positioned, the sunburst will be centered on the vertical line.

Take a second pushpin to make pinholes through the template and in the wood to mark all the hour lines. Make one pinhole at either end of each hour line. Lift the template again to check your progress and accuracy. When you finish marking the hour lines, remove the template and repeat the transfer on the lower face of the dial. Since the underside is for telling time during the winter, when the days are short, you only need to mark hours for as early as 6:00 A.M. and as late as 6:00 P.M.

The two pinholes for each line should align with the dial's center point. Using a metal-edged ruler for a guide, lightly pencil in all the hour lines. The hour lines are 15° apart around the circle. It is sufficient if the lines are no longer than the distance between the pinholes. Pencil in shorter lines midway between the hour lines to mark for half-hours.

In order to draw the roman numerals, you will need curved guidelines. You can make these by setting your draftsman's compass for a radius of 2" and, using the dial's center point for the compass needle, drawing a light circle on the wood. Make another circle with a 2¼" radius. Pencil in the roman numerals lightly, using these two lines as the upper and lower parameters. Leave ample space between the numerals so that they mark each hour line clearly. Repeat this procedure for the lower face, but note that *the hour sequence is reversed.*

Use a similar method to draw curved guidelines for your motto on the upper face. Use radii of 3¾" and 4⅛". You don't have to draw complete circles; arcs over the sunburst design will suffice. The motto you burn in will be permanent, so give it lots of thought.

See the template for the positions of "P.M." and "A.M." At the bottom of the board, below the dial face, pencil in your initials and the date.

Preparing the Base Plate

The second piece of wood is for the base plate. Divide the wood exactly in half with a pencil line along the vertical axis. Measure up 4" along this line from the bottom of the board. This will be the notch for 48° latitude. Using the vertical line and the 48° marking to align the template and wood, locate the positions for the remaining notches (44°, 40°, 36°, 32°, 28°, and 24°) on the board with the carbon-paper transfer method.

Now remove the template and lay the brass rod diagonally across the wood; it will just fit inside the corners. Outline the rod in pencil. Use your matte

(Templates follow. Text for the third project resumes on page 21.)

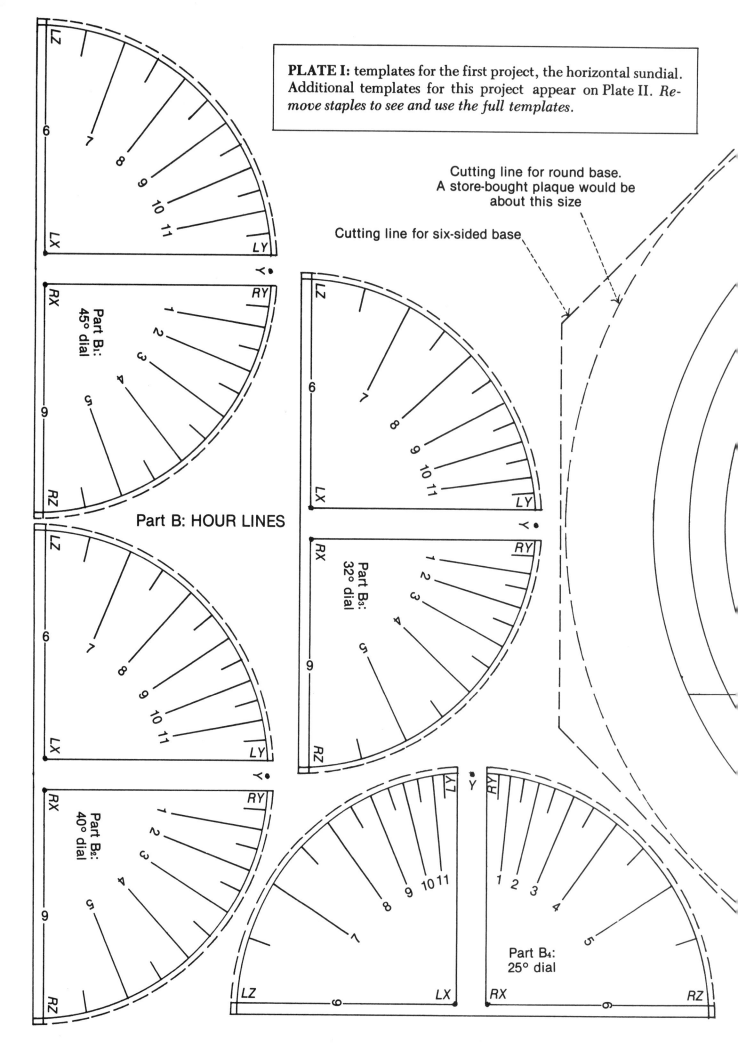

PLATE II: templates for the first and second projects, the horizontal and vertical sundials respectively. Additional templates for the vertical sundial appear on Plate III. Templates for the vertical sundial are calculated for use with a 7" x 10" board. *Remove staples to see and use the full templates.*

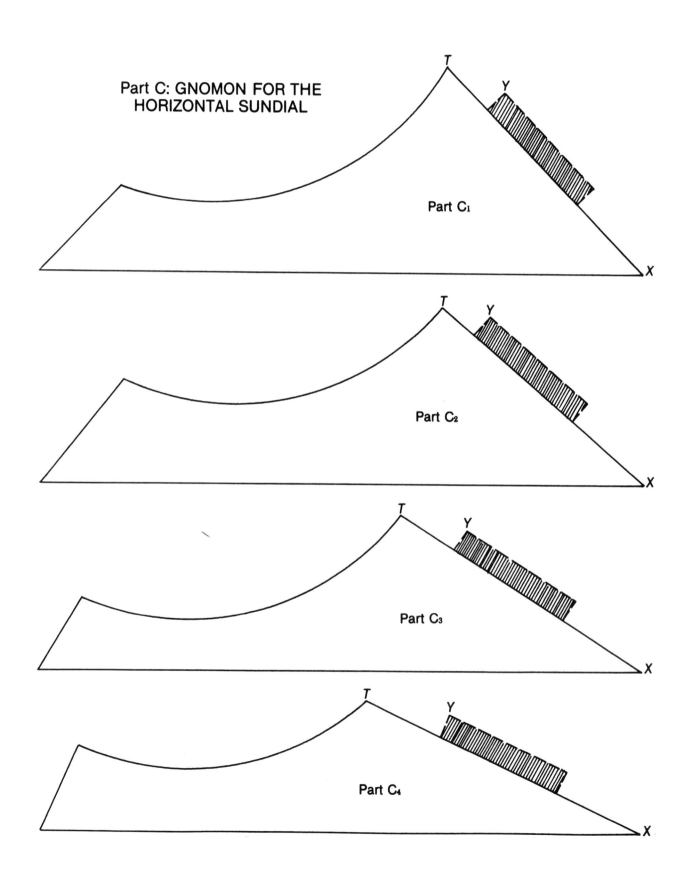

Part C: GNOMON FOR THE HORIZONTAL SUNDIAL

Part C_1

Part C_2

Part C_3

Part C_4

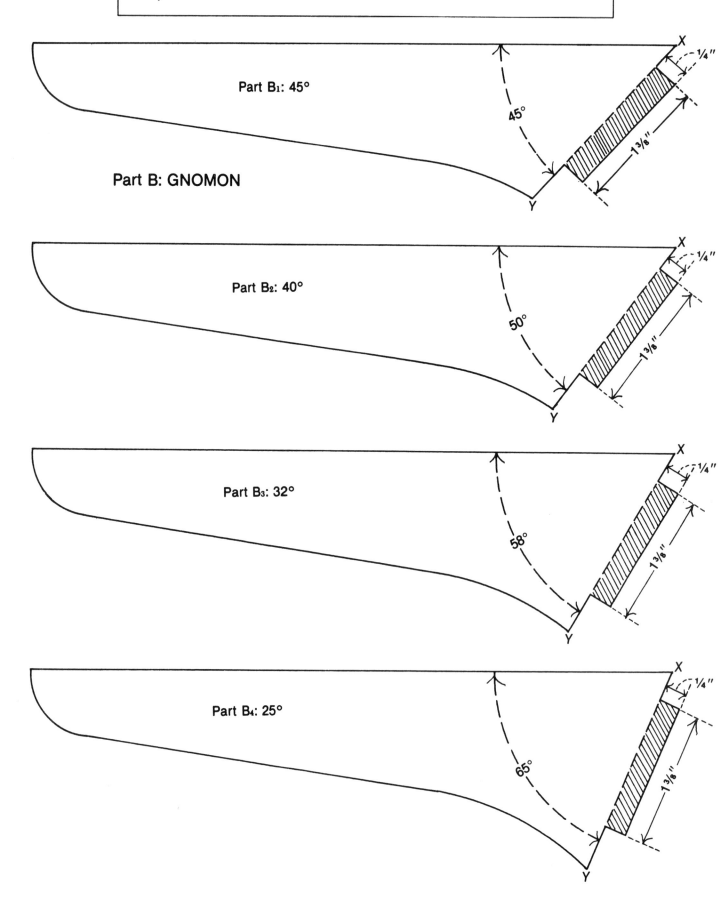

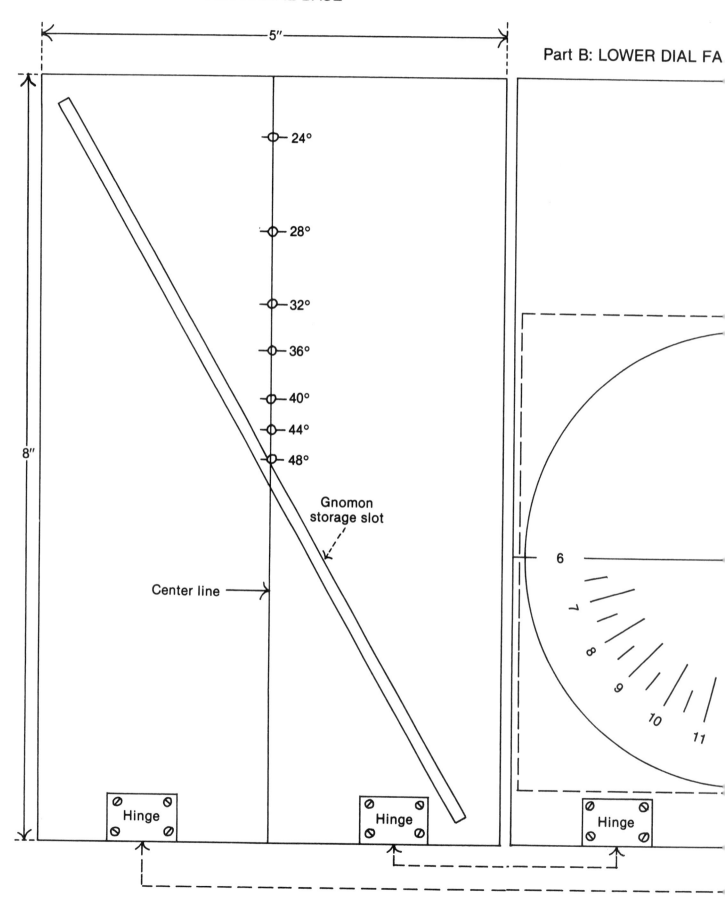

PLATE V: template for the fourth project, the diptych sundial, calculated for four latitudes. *Remove staples to see and use the full template.*

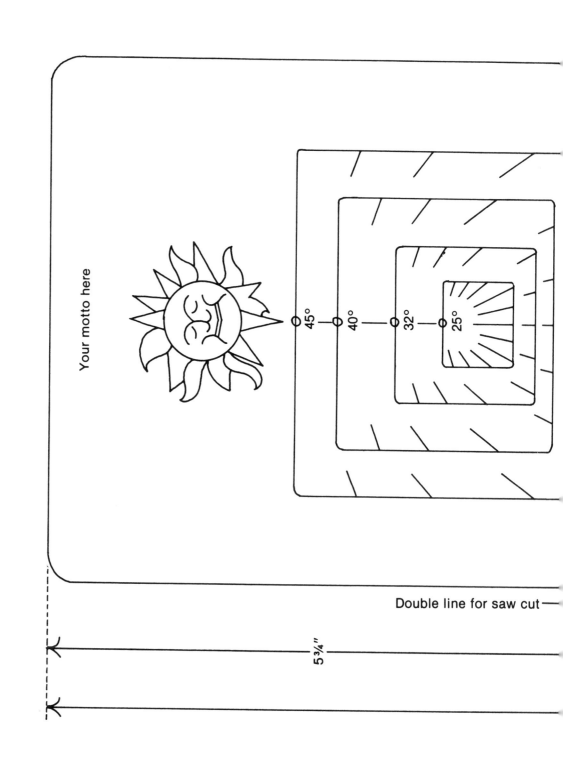

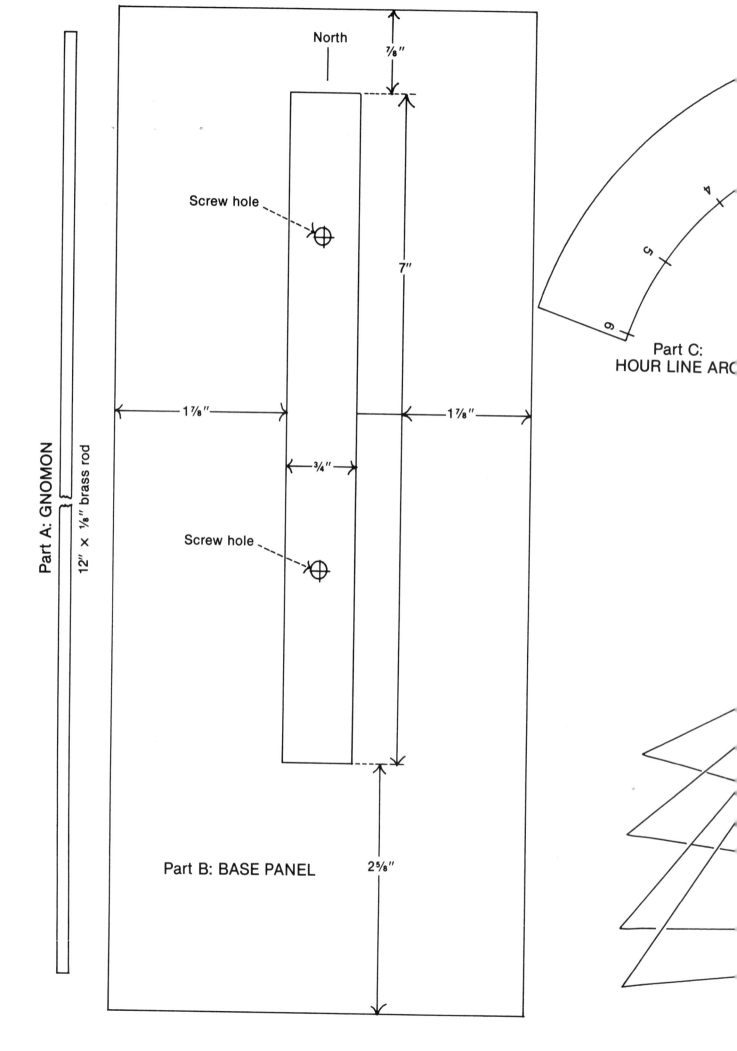

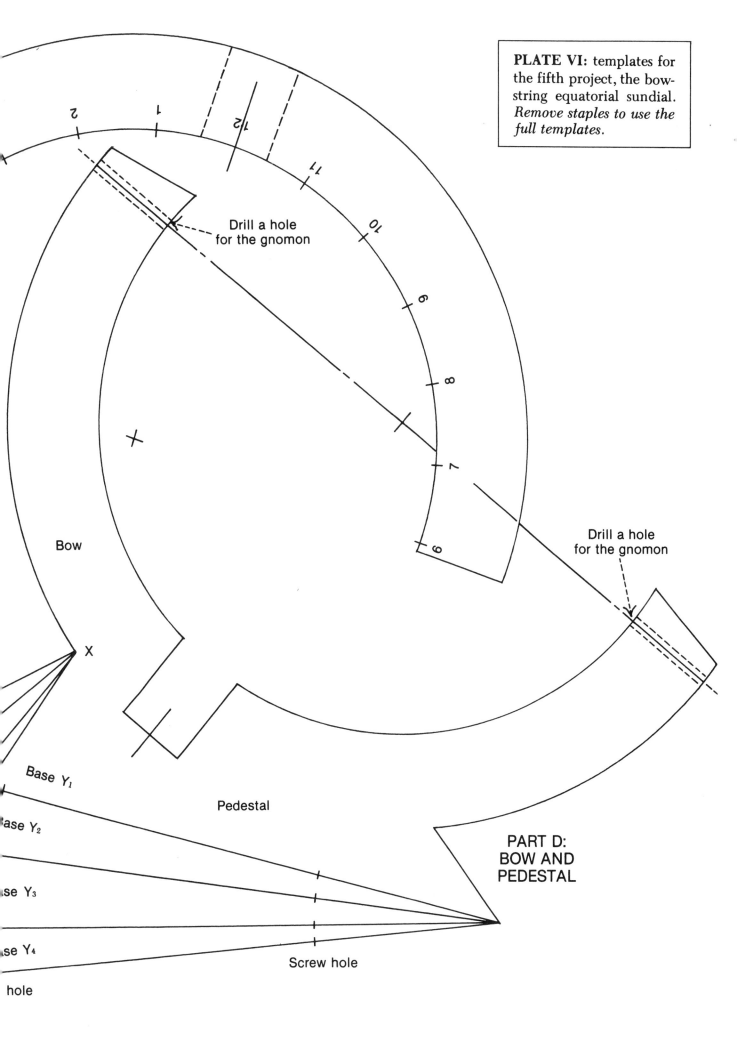

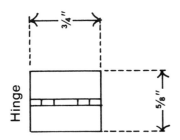
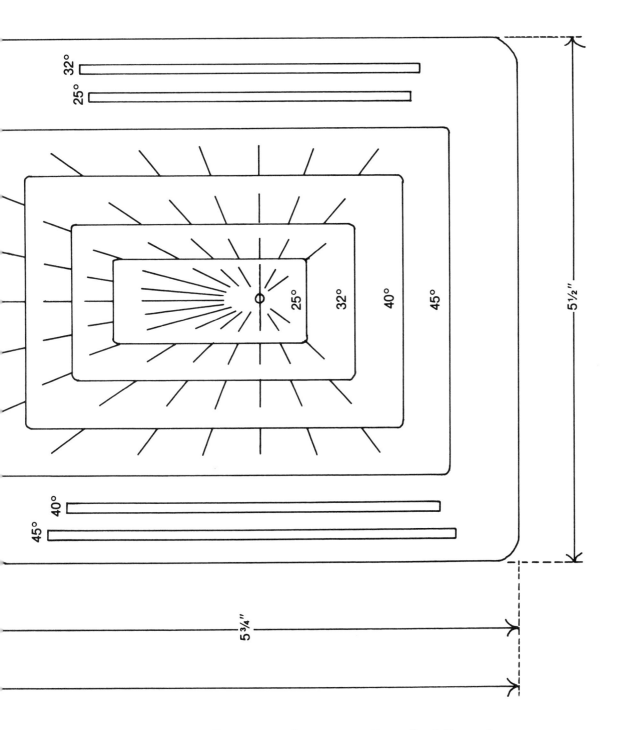

Remove staples to see and use the full templates.

PLATE IV: templates for the third project, the folding equatorial sundial, calculated for seven latitudes. *Remove staples to see and use the full templates.*

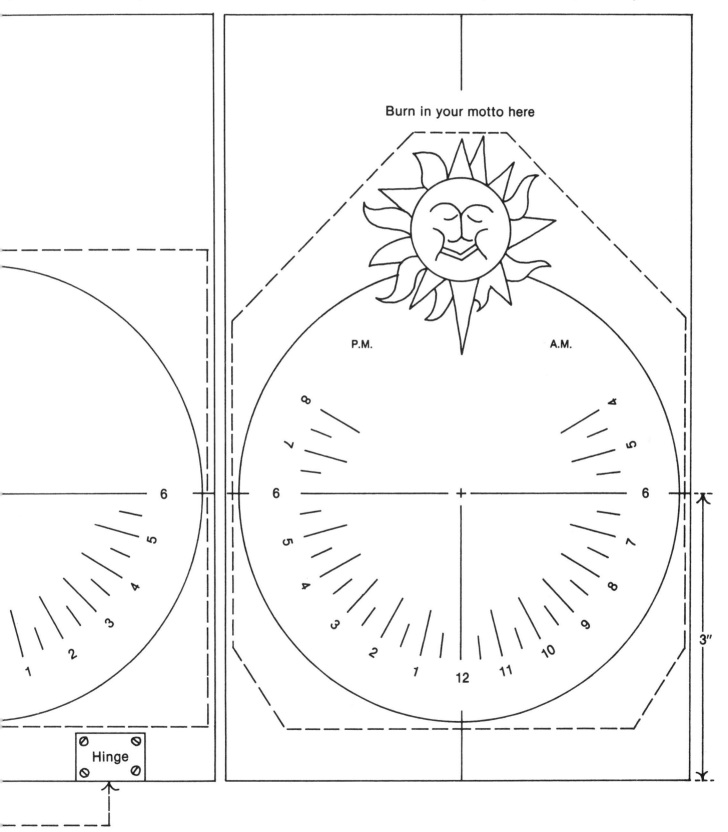

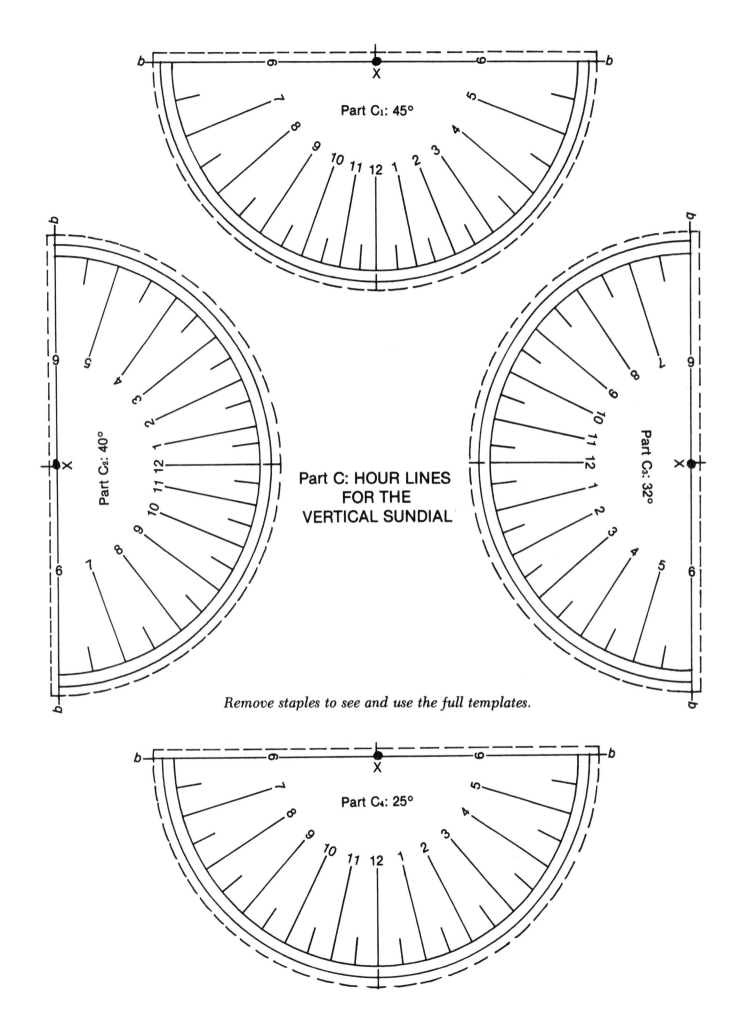

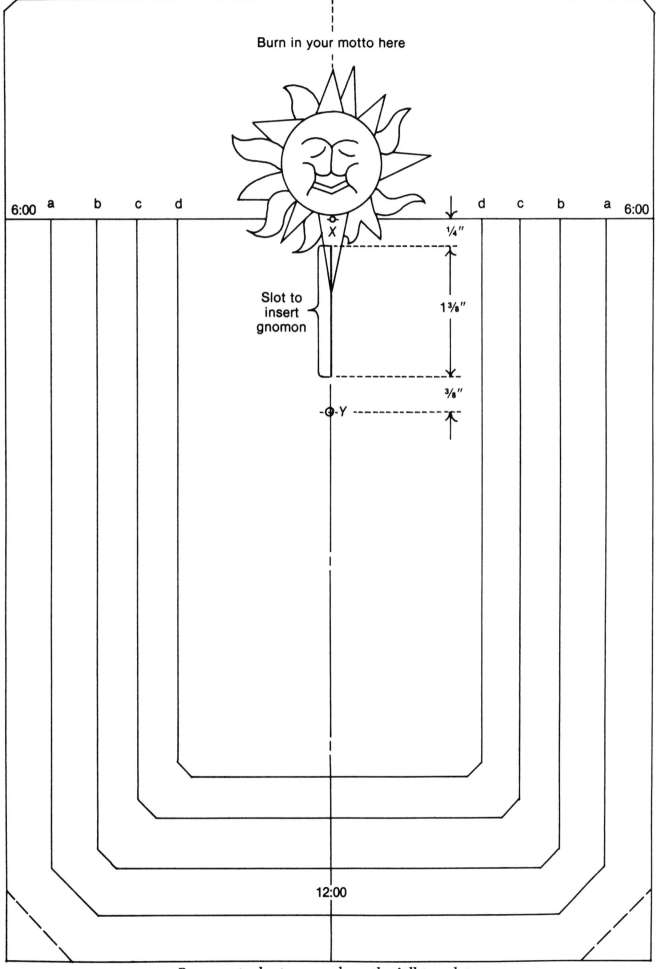

Part A: DIAL FACE FOR THE VERTICAL SUNDIAL

Remove staples to see and use the full templates.

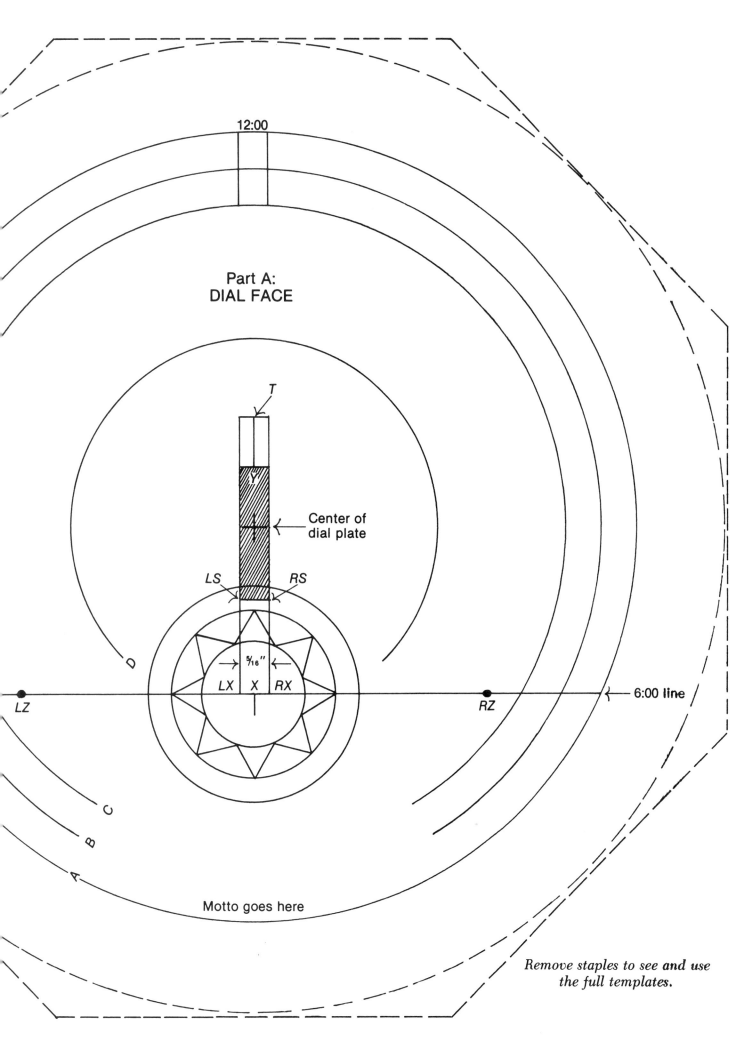

(Text for the third project resumes here.)

knife to cut along the outline to make a slot to store the gnomon. Make the slot fairly deep but not too wide or long: the gnomon must fit snugly. Remember to make the small latitude notches with your matte knife.

Joining the Plates

Place the two plates end to end and lay the hinges open on them (see Figure 12). The hinges' spines should be precisely on the line where the two boards meet. Outline the hinges with pencil neatly and exactly. With your matte knife slice firmly into the wood along the outlines. Remove the hinges and cut along the lines again, making your incisions slightly deeper than the thickness of the hinges. Chip the wood out cleanly and evenly to a depth equal to the metal thickness. The hinges should fit into their grooves perfectly, forming a smooth and even surface with the surrounding wood. Check again to be sure that the hinges' spines align perfectly over the line where the two boards meet. This is very important in order for the two panels, when folded face to face, to coincide exactly, with their edges flush all around. Press a pushpin through the screw holes to form starter holes for the screws. Screw in the hinges. When folded together the edges of the two panels should align perfectly.

Inserting the Gnomon

Before you can insert the gnomon, you must separate the two panels by removing the screws from one leaf of each hinge. Making the hole in which the gnomon is to be inserted requires particular care. Using the thinnest drill bit you have, drill a hole into the dial center, the point from which all the hour lines radiate. Be sure you hold the drill perfectly vertically, not leaning in any direction. The drill bit should come out precisely through the dial center on the other side of the panel. Now, using a ⅛" bit (the diameter of the gnomon), drill through the first hole again. The gnomon should slide into the hole—not easily: it should require a little forcing. When the gnomon is inserted, it should be at right angles to both sides of the dial panel. At this point you should remove the gnomon, reassemble the folding dial by rescrewing the hinges, and reinsert the gnomon. Push the rod through the dial until the lower end rests snugly in the latitude notch of your choice.

With the gnomon in place, your sundial is now set so that the gnomon forms an angle with the base panel equal to your latitude (see Figure 13). Math buffs will be interested to know that the angle between the two panels is the complement of the latitude—90° minus the latitude. This is so because the gnomon forms a right (90°) angle with the upper dial plate and every triangle has a total of 180°.

Burning in the Hour Lines and Details

You may be an old hand with an electric pen by now, but if you're still not confident with the tool, it's a good idea to practice some more with a piece of scrap wood. After your burn in the hour lines, don't forget the sunburst design, your motto, initials, and the date. Burn in the center line on the base panel, also; it will serve as a north-south line when you orient the dial with a magnetic compass. Now is also the time to add any personal distinguishing decorations, but don't overdo the ornamentation at the expense of the dial's simplicity.

Optional Magnetic Compass

Before you finish the dial, you may wish to add a small magnetic compass to the lower panel. The compass is for setting the dial; if you don't want to put a compass on the dial base, you will still have to use a compass to orient the dial to the north. Simple, inexpensive compasses may be purchased in Boy Scout supply stores, novelty shops, and the gift store of a museum with an astronomy exhibit. The compass must be very thin, much thinner than the wood base.

Place the compass exactly on the center line of the base panel between the gnomon slot and the hinges. Sink a hole as deep as the compass is thick. The diameters of the hole and the compass must be identical. With its north and south points coinciding

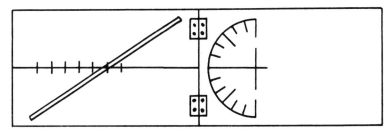

Figure 12. Aligning the plates.

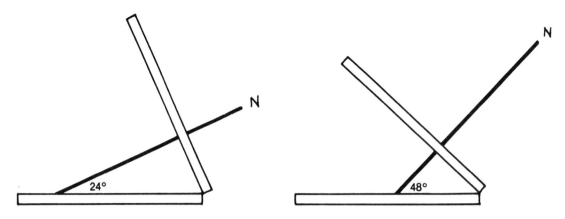

Figure 13. Two settings for the folding equatorial dial.

with the base panel center line, the compass should rest snugly in the hole. The recessed compass should not interfere with the folding of the dial.

Finishing

Remove the hinges once again and finish the two panels separately. Using medium sandpaper, smooth all the surfaces and sides of the two dial panels. If you handled your pencil lightly, all guidelines and finger marks should disappear quickly with the sanding. Round the corners of the panels with the sandpaper. Do the same to the edges. Then use fine sandpaper or very fine steel wool. With all pencil lines and blemishes removed and the wood smooth to the touch, the burned lines and details should stand out in pleasing contrast to the light wood.

Apply a very light stain (which will also serve as a wood filler) to every surface of the panels and rub them firmly with a cloth. The stain will intensify the brown color of the incised details and leave the wood surface with a dull, satin patina. Continue finishing with a couple of coats of clear, high-quality, spar varnish. When the first coat is thoroughly dry (without a trace of tackiness), rub the wood down lightly with extra fine steel wool. Wipe away the dust and apply a second coat of varnish. Rub the wood down with a cloth to get a smooth finish. Reset the hinges and insert the gnomon. Your folding equatorial sundial, set to your latitude, is now ready for the sun.

Setting the Dial

Let's begin at the beginning, with the dial folded and the gnomon stored in its slot. Unfold the dial and insert the gnomon into the dial's center hole. Push the gnomon through the hole until the gnomon's lower end touches the appropriate latitude notch. Make sure that the dial is perfectly level on its windowsill, table, or other surface that is fully exposed to the sun. Orient the dial with a magnetic compass so that the gnomon points north. Now you may read your dial: upper face for the summer, lower face for the winter. Be sure to consider your location's magnetic variation correction, as discussed in the previous projects.

FOURTH PROJECT

DIPTYCH SUNDIAL

Diptychs, tablets with two leaves folded together to protect writing on the inner surfaces, were used by the Greeks and Romans for recording commemorative, religious, and instructional information. Small, portable diptych sundials were usually made of ivory, wood, or metal. Diptych sundials were also made during later periods. Some museums exhibit exquisite European and Oriental examples from the fourteenth, fifteenth, and sixteenth centuries.

The diptych sundial is a simple combination of the horizontal and vertical dials. The lower half of the hinged unit is level like a horizontal dial; the upper half is upright like a vertical dial. The two leaves are held at right angles by the gnomon. The angle of the gnomon varies according to the latitude. The templates for this project (see Plate V) have been designed for 25°, 32°, 40°, and 45° latitude. There are four dial faces, one inside another, on each leaf. There is a single notch on the horizontal leaf at the midpoint of the 6:00 line for placing the gnomon. Four gnomon notches, one for each latitude, appear on the vertical leaf.

Materials

In addition to many of the tools and materials required for the preceding projects, you will need a single piece of wood 12″ × 5½″ × 3/16″. I recommend clear pine shelving stock. The brass rod for the gnomon should be about 16″ long, later you will cut it into four different gnomons, one for each latitude. Purchase a pair of ¾″ × ⅝″ solid brass hinges. To cut the wood into two leaves you will need a saw with fine teeth.

Transferring the Patterns

Use the carbon-paper transfer method. In order to align the templates properly with the wood, lightly draw a line down the center of the wood. Lay the template and carbon paper on the wood, with the 12:00 line exactly over the center line. Pin the template to the wood and trace the pattern carefully. Where applicable, use a pushpin as a ruler guide—for example, a pin at the intersection of the 6:00 and 12:00 lines will guide you as you draw all the hour lines. Do not forget to draw the lines for the four gnomon slots. If you perform the transfer carefully, the completed leaves should look exactly like the templates.

Incising and Cutting

In the hands of a skilled craftsman, an electric pen can be counted on for very fine, detailed work; but if you are relatively new to the electric pen, the closeness of many of the lines of the diptych sundial is a good reason to use another way to mark the wood. I recommend that you use a black ball-point pen. The fine point will make a clean, straight line. Use pushpins and a metal-edged ruler to guide the pen. Be careful not to press too hard: the pen may enter the grain of the wood and be deflected from its proper path.

Now that you have marked all the lines, cut the wood into the two leaves. Note that there are double lines on the pattern between the two leaves; the space between the lines is for the width of the saw. File the corners of the leaves until they look like the drawings. Gently rubbing with steel wool will remove finger marks and pencil lines without erasing the black pen lines. Use a matte knife to slice out the storage slots for the gnomons. Do not overdo the cutting. The gnomons should fit in the slots snugly.

Hinging the Leaves

The hinges will be attached in *reverse position* so that the upper diptych panel, when open, will be held at a right angle to the lower panel. Each hinge spine must rest *between* the panels, and the hinge leaves should lie flat against the thin panel ends with the screw hole bevels on the *underside* (see Figure 16). (This means that the screw heads will *not* take advantage of the screw hole bevels.) You

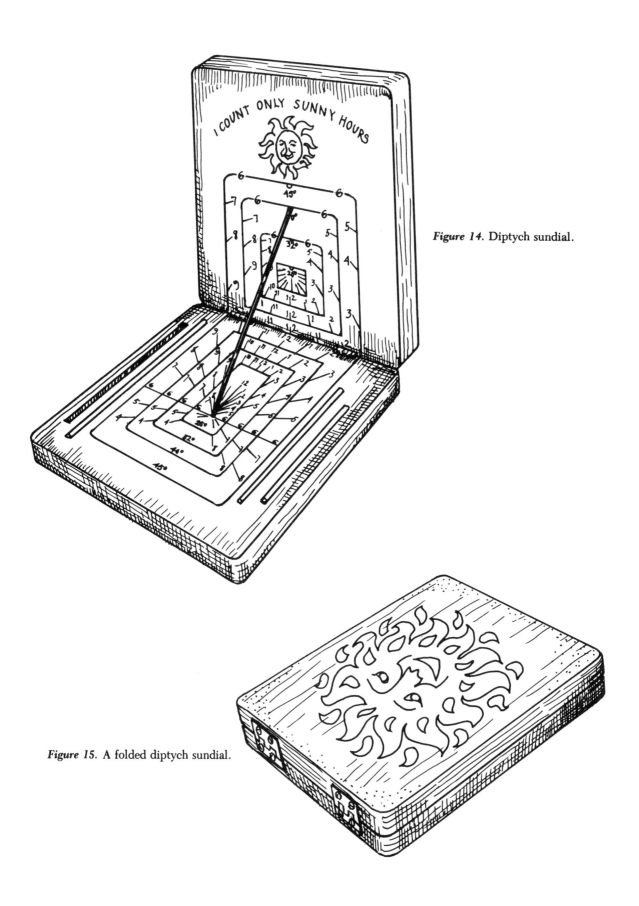

Figure 14. Diptych sundial.

Figure 15. A folded diptych sundial.

24 *Easy-to-Make Wooden Sundials*

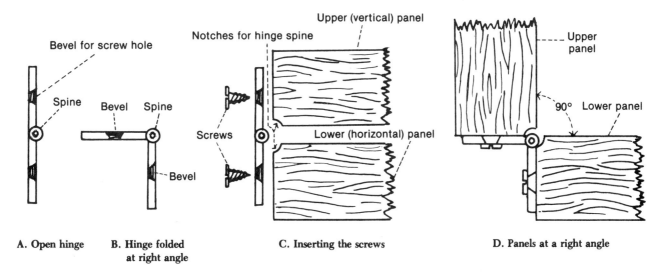

Figure 16. Hinging panels in reverse position.

may want to experiment with the hinges before attaching them to the wood. Instead of folding the hinge closed, turn one leaf back. You'll see that it does not fold all the way back but remains at a right angle, the position of the open diptych.

Place the two panels *face to face* and hold them together in position with a long piece of tape. Open the hinges and place them against the edges of the wood, with the hinges' spines toward the wood. You will see that the spines get in the way of the hinges so that they can't quite lie flat. Outline the hinges with pencil. Then remove the hinges and make notches in the panel ends with the matte knife; the notches are cut to accommodate the thickness of the spines. When the notches are completed, the hinges will be able to lie flat against the wood. Use a pushpin to make starter holes for the screws. Use brass screws to attach the wooden panels. If the diptych is hinged properly, the two panels should fold together face to face; when open, the panels should be at a right angle.

The Gnomons

The proper lengths for the four gnomons are shown on the template. Cut your brass rod to the correct lengths. Make a single notch on the lower dial where the 6:00 and 12:00 lines meet. Four notches are required on the upper panel; they go where the center line of the panel intersects each of the four dial faces. Note that since the dial faces are rectangular, the center line intersects each face at two points. Use only the topmost intersection for each face when making the notches.

Test each gnomon by placing it in the proper notches. For example, the 32° gnomon goes into the only notch there is on the lower (horizontal) panel, and into the 32° notch on the upper. Regardless of the latitude the diptych should always be open at the hinges, making a perfectly erect right (90°) angle.

Finishing

Before finishing the wood it's a good idea to test the dial in the sun. Remember to orient the dial properly, pointing the gnomon at the polestar. The tested dial should be sanded, stained, rubbed with steel wool, and coated with protective varnish in accordance with the instructions provided for previous projects.

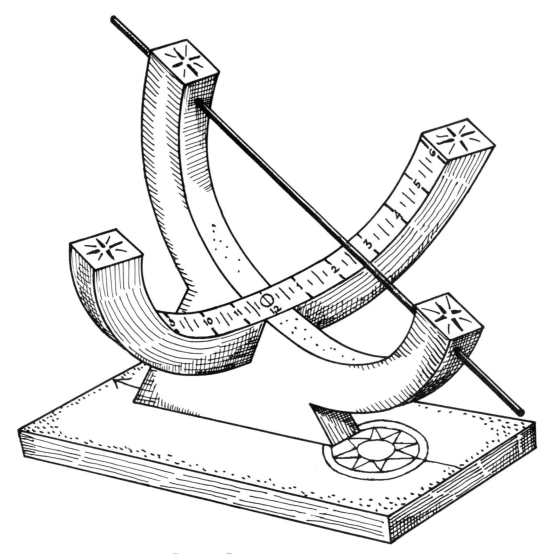

Figure 17. Bowstring equatorial sundial.

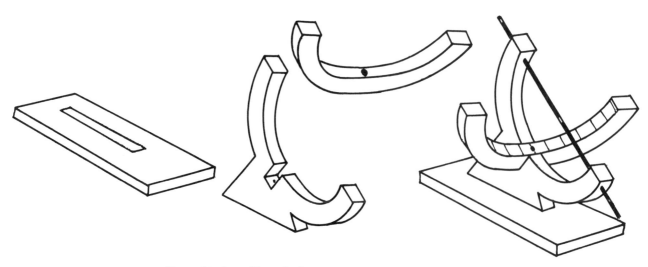

Figure 18. Assembling the bowstring equatorial sundial.

26 *Easy-to-Make Wooden Sundials*

FIFTH PROJECT

BOWSTRING EQUATORIAL SUNDIAL

The bowstring equatorial sundial is, in principle, a variation of the equatorial dial of the third project. The bowstring dial has four parts; you have templates for all of them (Plate VI). The template marked Part A is for the gnomon (the string of the bow). Make the gnomon from a brass rod 12″ long and ⅛″ in diameter. Using brass is important because it is nonmagnetic. Part B is for the base. Part C is the arc with hour lines—the part of the dial that you will read. Part C is designed to fit, concave side up, into a notch in Part D, the bow and pedestal section. Part D (the bow) is the only part of the sundial that varies according to latitude; the plans for Y_1, Y_2, Y_3, and Y_4 are for 25°, 32°, 40°, and 45° respectively. Part D fits into the base (Part B).

Parts B, C, and D may all be cut from the same piece of wood. Use first-grade 12″ clear white shelving pine; this should be about ¾″ thick. An 18″ board will do, but I recommend a 20″ one to make it easier to grip and cut the wood.

Procedure

You may transfer the patterns to the wood with carbon paper, but I prefer the following pushpin method. Pin the templates to the wood. Then take one pin and push it through a template and into the wood at the important points of the template. These points are at corners of the drawings, at crosses where you will later place the point of your draftsman's compass, and at points where you will drill and sink screw holes. Unpin the template on the left side (*not* both sides because you don't want the template to change position). Using a pencil and ruler, connect the appropriate dots. Use a compass to draw the arcs. When you are finished with the left side of the template, repin it and work on the right side. The completed board should look like the template and be ready for cutting into the three wooden components.

On the base (Part B) pencil and score the lines that indicate the position for mounting the bow (Part D). Drill the two screw holes.

Cut out the bow, first making sure that you have chosen the correct pattern for your latitude. Notching the bow where the hour line arc fits in requires care to make it perfectly square. First use a saw to cut both sides of the notch to the proper depth. Then saw several more cuts between the first two. At this stage the notch should look like the teeth of a comb. Use a screwdriver between the teeth to push and twist away the slivers of wood. *Only if necessary* apply a wood file to the sides and bottom of the notch to insure a perfect fit when you later insert the hour line arc.

Note that the bow is slightly longer than a semicircle. The extra length is for the holes where the gnomon will be inserted. Drill the holes with the same diameter as the metal rod. Follow the template for the *exact* position and direction of these holes.

Cut out the hour line arc, lay it flat on the pattern, and mark off the hour lines. The center of the arc will be for 12:00. When the arc is in place and the gnomon points north, the shadows for the morning hours will appear on the left (west, because the sun is in the east before noon); afternoon hours will appear on the right. Subdivide the hour divisions by eye as you please—into halves, quarters, or thirds. Burn in the lines with the electric pen.

Now that all the part have been cut out, they are ready for a test assembly. The hour line arc should fit into the bow at a right angle; use a square piece of stiff cardboard to check this. You may have to use a wood file to improve the fit of the pieces. To secure Parts C and D together, drill a hole through the 12:00 line and into the bow at the bottom of the notch. Using nonmagnetic brass screws, assemble all the pieces and check the dial for accuracy (see Figure 18.

Once the parts are all shaped and fitted accurately, disassemble the dial for finishing. Follow the decorative and finishing instructions as outlined earlier. Orient the bowstring dial to the north in the manner described for the previous projects.

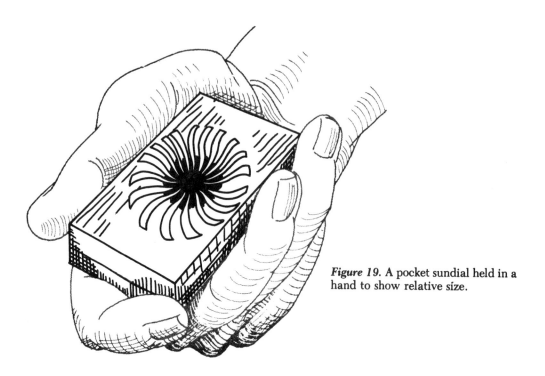

Figure 19. A pocket sundial held in a hand to show relative size.

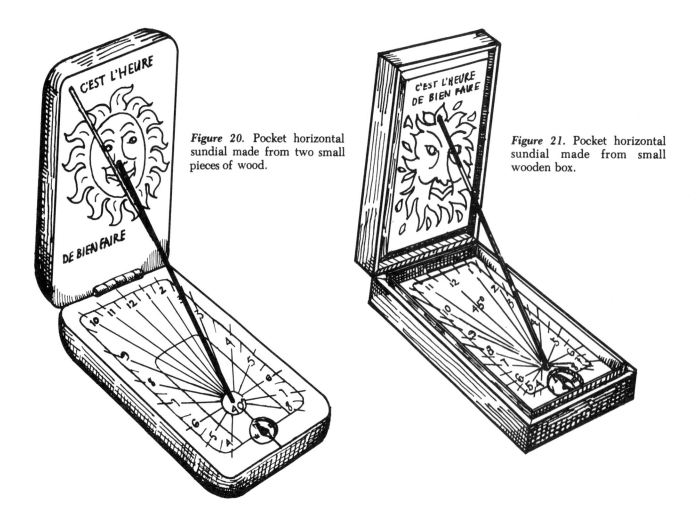

Figure 20. Pocket horizontal sundial made from two small pieces of wood.

Figure 21. Pocket horizontal sundial made from small wooden box.

28 *Easy-to-Make Wooden Sundials*

DESIGN YOUR OWN POCKET SUNDIAL

The innovative craftsmanship of the old dialists is evident in the variety of clever miniature portable sundials that survive today. At this point in the book perhaps you can apply your knowledge and skill to making a small dial of your own design. For example, with the instructions, templates, and transfer steps used for the horizontal sundial, you can make many different small dials. I made one from a small 1¾" × 3" × ¾" wood box purchased in an arts and crafts supply shop. Another was made from two pieces of ⅜" thick scrap wood, a small brass hinge, and a short, thin brass rod for the gnomon—a simple, folding horizontal dial.

Let's make a folding horizontal pocket dial. You can make it any size, shape, or design you choose. For this example we will use two matching pieces of scrap wood, each about 4" × 3" × ⅜". When it is hinged and folded along the 3" edge, the dial slips easily into a pocket.

From the horizontal plan choose the dial face with the hour angles and lines of your latitude. Note that the dial face plan has double 12:00 lines to accommodate the thickness of a wood gnomon. The pocket dial now being considered, however, will have a thin brass gnomon of negligible thickness and thus need only a single 12:00 line. The transfer of the dial face to the small wood panel takes two steps. First trace the dial face onto a sheet of paper by slipping carbon paper between the template and the paper. Keep in mind that the tracing will be no larger than the wood panel. The two sides, morning and afternoon, will be next to each other, with a single, common 12:00 line. After you trace the morning (left side) and afternoon (right side) hours, draw the horizontal 6:00 line and extend the 7:00 and 8:00 morning hours to form the 7:00 and 8:00 afternoon hours. Make similar extensions of the 4:00 and 5:00 afternoon hours to form the 4:00 and 5:00 morning hours. All these extensions are short lines below the 6:00 horizontal line. The 12:00 vertical line crosses the 6:00 line at a right angle and extends slightly below the 6:00 line.

Next, draw a rectangle the size of your wood panel (3" × 4") around the transferred dial face. The position of the dial face 12:00 line should be exactly on the vertical center line of the 3" × 4" rectangle, and the 6:00 line should be about one-third of the way *up* from the bottom of the rectangle. On the wood panel lightly pencil a vertical line exactly down the center. Also pencil the horizontal 6:00 line one-third of the way *up* from the bottom (3") edge. The center of the dial face is where these lines cross at a right angle.

Now place the new template tracing on the wood panel with the 12:00 and 6:00 lines properly aligned. By means of the pushpin method described in previous projects, transfer the hour line markings to the wood panel. Remove the paper template and with your metal-edged ruler and black ball-point pen incise the hour lines onto the wood.

Before finishing and hinging the horizontal dial face panel, the vertical wood panel must be prepared properly. This step is important; it determines where the gnomon set at the latitude angle terminates in a notch in the center line of the vertical panel (see Figure 22). The list of multiplying factors (see Table I) shows where to put the notch. The factors are based on trigonometric functions of the latitude angles and are not discussed in this book. The listed factors are precalculated for simplicity's sake.

For our example we are making a 4" long pocket

		For our example	
		Horizontal (lower) panel	Vertical (upper) panel
Latitude (°)	Factor	4"	4"
45	1	2.75	2.75
40	.839	2.75	2.3
32	.624	2.75	1.7
25	.466	2.75	1.3

Table I. Factors for designing a pocket sundial.

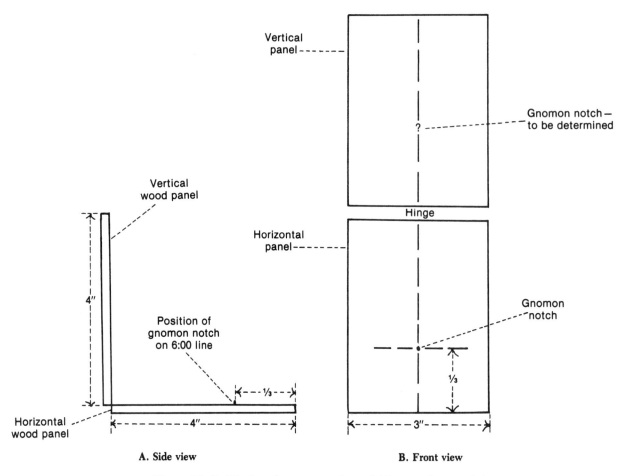

Figure 22. Positioning the gnomon for a folding pocket sundial.

dial for 25° latitude. Let's place the 6:00 horizontal line 2¾" *down* from the hinged edge of the lower panel. The latitude factor for 25° is .466. Multiply the 2.75 (2¾)-inch dimension by the factor .466; you will get 1.3 inches. This will be the distance *up* from the hinged edge for the notch on the upper panel. Convert this figure to eighths or sixteenths of an inch; it will be close enough for the accuracy we wish to achieve. Notch the spot on the upper panel. This will be the upper terminating point of the gnomon. When the hinged panels are in proper position, set at a right angle, the inserted gnomon will be resting at an angle of 25° and pointing to the North Star (at 25° latitude).

For another example, you could make a 5" panel for 40° latitude with the 6:00 line 3" from the hinged edge. Multiply 3" by the factor .839, which will give you 2.5". Thus the notch will be 2.5" *up* from the hinged edge on the upper panel. When set in the two notches, the gnomon will be at a 40° angle—the setting for 40° latitude.

Now let's make the gnomon for the 4" dial. You will need a thin, rigid, brass (nonmagnetic) rod. Draw a right angle on a piece of paper. The notch point on the lower panel of our example is 2¾" from the hinge. Mark the lower leg of the right angle 2¾" from the corner. The vertical notch for our example is 1⅓" up from the hinge. Mark the upper leg of the right angle 1⅓" from the corner. Between the two points on the right angle draw a straight diagonal line. The length of this diagonal will be the length of the gnomon. Cut the rod longer, however, to allow for inserting the ends into the notches.

Complete the two panels by adding some border decorations with your ball-point pen or wood-burning tool and laying on a motif on the vertical panel. Of course, add a pithy motto. Diagonally across the vertical panel cut in a straight slot for storing the gnomon when the dial is folded. Decorate the top surface of the folded dial. Finally, lightly sand, stain, and varnish the wood. Affix the hinge in the manner described in the plan for the diptych sundial. Above or below the 6:00 line carve out a hole for a tiny magnetic compass.

CORRECTING YOUR SUNDIAL

If you have paid very close attention to the instructions in this book and constructed your sundial carefully, why do you have to correct the reading? For one thing, the speed of the sun (rather the speed of the earth) varies at different times of year; hence all sundial readings necessarily involve averaging the sun's speed over the course of a year. For another, as mentioned earlier, most of us are content to live by standard time; thus a sundial, which shows sun time, must (or can) be corrected to give standard time.

Correcting for the Equation of Time

Since the earth leans away from its vertical axis as it orbits the sun, and since the speed of its orbiting varies from month to month, the equation of time has been developed to determine the sun time for any given date. For our purposes it is not necessary to delve into the complex celestial mechanics and derive difficult mathematical formulas; all you need do is consult Figure 23, which provides a handy correction for the equation of time. The chart's curve plots how slow or fast your sundial is at any time of the year. Sun time can differ as much as 16 minutes with a clock located on the standard meridian of longitude. Without the equation of time your dial is only absolutely accurate four times a year—about April 16, June 14, September 1, and December 25. (For a clock that is not located on a standard meridian of longitude, see the section Longitude Correction.)

To make your correction, read the dial, then follow the curve on the chart to the proper month and approximate day. When you have located the correct point, look at the top of the chart, where you will see how many minutes to add or substract from your reading. For the sake of easy reference, you may want to incise the chart on one panel of the sundial itself.

Longitude Correction

The equation of time will help you go from sun time to mean time, but it won't give you the time you hear on the radio, standard time. Standard time is the time over a broad area, whereas sun time is for only one meridian of longitude. There are twenty-four standard time zones, one for each of the twenty-four hours it takes the earth to rotate once. There are 360° in a circle; thus each zone is 15°, and each degree represents 4 minutes of time. By international convention, the zones take their reference from Greenwich, England (0° longitude). The standard zone meridians lie midway between the east and west bounds of their respective zones. Thus 15° west of Greenwich is the meridian of zone 1, 30° west of Greenwich is the meridian of zone 2, and so on. The continental United States has four zones (5–8) (see Figure 24 and the List of Selected Cities).

Since standard time averages mean times over a broad (15°) area, to go from your sundial reading (already corrected for the equation of time) to standard time, you have to know where your exact location is in relation to your time zone's standard meridian of longitude. For example, as you can see in Figure 24, New York City is in zone 5. There are 15° per zone; therefore New York's standard meridian of longitude is 75° (15° × 5). New York's exact longitude, however, is 73°50′; thus you must correct your dial for the difference between 75° and 73°50′—1°10′ or roughly 4 minutes of time. For another example, Topeka, Kansas, is in zone 6; its standard meridian is therefore 90° longitude; but its actual meridian is 95°41′. Thus dialists in Topeka must correct their readings for 5°41′—the difference between 90° and 95°41′—or 23 minutes of time.

There is one other thing to keep in mind when you make your longitude correction. Since the standard meridian of a time zone cuts its zone in half, half of the zone lies east of the standard meridian, half west. The sun appears to move from east to west; thus if your dial is east of the standard meridian it is said to be fast (because the sun gets there

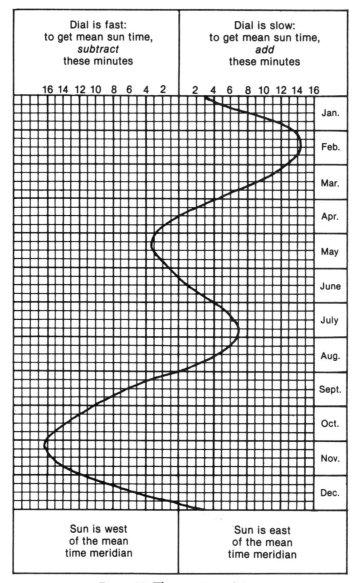

Figure 23. The equation of time.

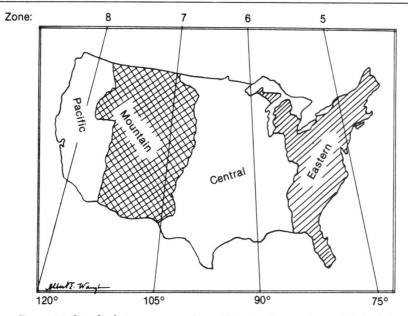

Reprinted from Albert E. Waugh, *Sundials: Their Theory and Construction* (New York: Dover Publications, 1973), p. 14.

Figure 24. Standard time zones and meridians in the continental United States.

	Horizontal Sundial				Vertical Sundial			
	25°	32°	40°	45°	25°	32°	40°	45°
12:30 or 11:30	3°11'	3°59'	4°50'	5°19'	6°48'	6°22'	5°45'	5°19'
11:00 or 1:00	6°27'	8°05'	9°46'	10°44'	13°39'	12°48'	11°36'	10°44'
10:30 or 1:30	9°56'	12°23'	14°55'	16°20'	20°35'	19°21'	17°36'	16°20'
10:00 or 2:00	13°43'	17°01'	20°22'	22°12'	27°37'	26°05'	23°51'	22°12'
9:30 or 2:30	17°58'	22°08'	26°15'	28°29'	34°49'	33°03'	30°27'	28°29'
9:00 or 3:00	22°55'	27°55'	32°44'	35°16'	42°11'	40°18'	37°27'	35°16'
8:30 or 3:30	28°51'	34°38'	39°57'	42°40'	49°45'	47°52'	44°57'	42°40'
8:00 or 4:00	36°12'	42°33'	48°04'	50°46'	57°30'	55°45'	53°00'	50°46'
7:30 or 4:30	45°35'	51°59'	57°12'	59°38'	65°27'	63°58'	61°36'	59°38'
7:00 or 5:00	57°38'	63°11'	67°22'	69°15'	73°32'	72°28'	70°43'	69°15'
6:30 or 5:30	72°42'	76°03'	78°26'	79°27'	81°44'	81°11'	80°15'	79°27'
6:00	90°00'	90°00'	90°00'	90°00'	90°00'	90°00'	90°00'	90°00'

Table II. Chart of hour angles for four different latitudes.

before it reaches the standard meridian), and if it is west of the meridian, it is said to be slow (because the sun gets there after it reaches the standard meridian). You must substract minutes from a fast dial, and add minutes to a slow one, to complete your longitude correction. New York City is east of the standard meridian, therefore you must substract 4 minutes from your reading. Topeka is west of its standard meridian, therefore Topeka dialists must add 23 minutes of time.

Correcting for Magnetic Variation

Strictly speaking, it's not your sundial but your magnetic compass that must be corrected for magnetic variation, but since you will probably use the compass to orient your gnomon to true north, it's important that you account for magnetic variation when taking a sundial reading. To state the problem briefly, the earth's magnetic and geographical poles do not exactly coincide; hence a compass that points to magnetic north does not point to true north (except when the compass is located in an area where the magnetic variation is 0°). The variation is measured in degrees east or west of true north. See the List of Selected Cities, which has a column for magnetic variation.

If you are making a sundial in or around Hartford, Connecticut, for example, you have to correct for a local magnetic variation of 14°W. This means that your compass needle is being pulled 14°W of true north. With your compass resting on your sundial base, and with the compass needle and the gnomon in perfect alignment, turn the dial base until the compass reads 14°W (that is, 360° minus 14°, or 346°). Now both compass and sundial point to true north.

For another example, the magnetic variation in Denver is 12°E, which means that you would have to turn your dial base (with the compass on it) until the compass needle points to 12°. If you live in Chicago, no correction is necessary, since Chicago's magnetic variation is just about 0°.

You can round off to whole degrees when correcting for magnetic variation—the approximations will be accurate enough for the sundial projects in this book. If you have a superior compass, however, there is no reason why you shouldn't make your magnetic variation correction as accurate as possible.

Verifying the hour angles

Table II shows all the hour angles for the horizontal and vertical sundial projects; you can use the values on the chart and a protractor to check the accuracy of your dial. Note that there are four columns on the chart, one for each of the four latitudes for which the templates are designed. To use the chart, first locate the part you need—for either the horizontal or the vertical sundial. Then locate the proper column for your latitude. Say you made a horizontal dial for 32° latitude. The 12:00 line is 0°00' on your protractor. The angle for the 12:30 (or the 11:30) marking should be 3°59' from the 12:00 line. (It's okay to round off to whole degrees.) The angle for 11:00 (or 1:00) is 8°05', the angle for 10:30 (or 1:30) is 12°23', and so forth. Since the 6:00 and 12:00 lines are always at right angles to one another for both the horizontal and the vertical dials, the angle for the 6:00 line is 90°, regardless of the latitude setting for your particular sundial.

The chart of hour angles can also be used by advanced dialists who want to design their own horizontal or vertical sundials.

LIST OF SELECTED CITIES

	Latitude (North)	Longitude (West)	Time Zone	Add/ Substract Minutes	Magnetic Variation (°) (East/ West)
UNITED STATES					
Alabama					
Birmingham	33°30'	86°55'	6	−12	−1E
Mobile	30°40'	88°05'	6	−8	2.2E
Arizona					
Phoenix	33°30'	112°03'	7	+28	13E
Arkansas					
Little Rock	34°42'	92°17'	6	+9	4.5E
California					
Los Angeles	34°00'	118°15'	8	−8	14.2E
Sacramento	38°32'	121°30'	8	+6	16.5E
San Francisco	37°45'	122°27'	8	+9	16.4E
San Diego	32°45'	117°10'	8	−12	13.7E
Colorado					
Denver	39°45'	105°00'	7	0	12E
Connecticut					
Hartford	41°45'	72°42'	5	−7	14W
Delaware					
Wilmington	39°46'	75°31'	5	+2	10.5W
Florida					
Key West	24°34'	81°48'	5	+27	.7W
Jacksonville	30°20'	81°40'	5	+26	2.5W
Miami	25°45'	80°15'	5	+21	2.2W
Tallahassee	30°26'	84°19'	5	+37	.5W
Tampa	27°58'	82°38'	5	+30	1W
Georgia					
Atlanta	33°45'	84°23'	5	+36	1.2W
Savannah	32°04'	81°07'	5	+24	3.4W
Idaho					
Boise	43°38'	116°12'	7	+45	17.5W
Pocatello	42°53'	112°26'	7	+28	16E
Illinois					
Chicago	41°50'	87°45'	6	−9	.5W
Peoria	40°43'	89°38'	6	−2	1.6E
Springfield	39°49'	89°39'	6	−2	2E
Indiana					
Evansville	38°00'	87°33'	6	−10	.5E
Fort Wayne	41°05'	85°08'	6	−20	2.6W
Indianapolis	39°45'	86°10'	6	−16	1.2W
South Bend	41°40'	86°15'	6	−15	1.7W
Iowa					
Des Moines	41°35'	93°35'	6	+14	4.7E
Kansas					
Topeka	39°02'	95°41'	6	+23	6.5E
Wichita	37°43'	97°20'	6	+29	7.5E
Kentucky					
Louisville	38°13'	85°48'	6	−17	1E
Paducah	37°03'	88°36'	6	−5	1.5E
Louisiana					
Baton Rouge	30°30'	91°10'	6	+5	4.3E
New Orleans	30°00'	90°03'	6	0	3.7E
Shreveport	32°30'	93°46'	6	'15	5.5E

	Latitude (North)	Longitude (West)	Time Zone	Add/ Substract Minutes	Magnetic Variation (°) (East/ West)
Maine					
Augusta	44°17'	69°50'	5	−21	18W
Portland	43°41'	70°18'	5	−19	17W
Presque Isle	46°42'	68°01'	5	−28	18W
Maryland					
Baltimore	39°18'	76°38'	5	+6	9.5W
Massachusetts					
Boston	42°20'	71°05'	5	−16	15.7W
Pittsfield	42°27'	73°15'	5	−7	14W
Worcester	42°17'	71°48'	5	−13	15W
Michigan					
Detroit	42°23'	83°05'	5	+32	6.7W
Traverse City	44°46'	85°38'	5	+42	4.3W
Minnesota					
Duluth	46°45'	92°10'	6	+9	2.5E
Minneapolis	45°00'	93°15'	6	+13	4E
Mississippi					
Jackson	32°20'	90°11'	6	+1	3.5E
Missouri					
Kansas City	39°02'	94°33'	6	+18	5.7E
St. Louis	38°40'	90°15'	6	+1	2.5E
Springfield	37°11'	93°19'	6	+13	5E
Montana					
Billings	45°47'	108°30'	6	+14	15.4E
Great Falls	47°30'	111°16'	6	+25	17.4E
Nebraska					
Omaha	41°15'	96°00'	6	+24	6.5E

	Latitude (North)	Longitude (West)	Time Zone	Add/ Substract Minutes	Magnetic Variation (°) (East/ West)
Nevada					
Las Vegas	36°10'	115°10'	8	−19	14.5E
Reno	39°32'	119°49'	8	−3	16.6E
New Hampshire					
Berlin	44°27'	71°13'	5	−13	17W
Concord	43°13'	71°34'	5	−14	15.8W
New Jersey					
Atlantic City	39°23'	74°27'	5	−3	11.4W
Newark	40°44'	74°11'	5	−2	12.2W
Trenton	40°15'	74°43'	5	−1	11.5W
New Mexico					
Albuquerque	35°05'	106°38'	7	+6	11.8E
Carlsbad	32°25'	104°14'	7	−3	10.4E
New York					
Albany	42°40'	73°49'	5	−4	13.7W
Buffalo	42°52'	78°55'	5	+16	9W
New York	40°40'	73°50'	5	−4	12.3W
Plattsburg	44°42'	73°29'	5	−6	15.2W
Port Jefferson, L.I.	40°57'	73°04'	5	−7	13W
North Carolina					
Charlotte	35°03'	80°50'	5	+23	4.4W
Raleigh	35°46'	78°39'	5	+14	6.2W
North Dakota					
Bismarck	46°50'	100°48'	6	+41	10.3E
Fargo	46°52'	96°49'	6	+27	7E
Ohio					
Cincinnati	39°10'	84°30'	5	+39	2.8W
Cleveland	41°30'	81°41'	5	+27	6W
Columbus	39°59'	83°03'	5	+24	4.2W

List of Selected Cities 35

	Latitude (North)	Longitude (West)	Time Zone	Add/Substract Minutes	Magnetic Variation (°) (East/West)
Oklahoma					
Oklahoma City	35°28'	97°33'	5	+30	7.5E
Tulsa	36°07'	95°58'	5	+24	6.6E
Oregon					
Eugene	44°03'	123°04'	8	+12	19.4E
Pendleton	45°40'	118°46'	8	−5	19.2E
Portland	45°32'	122°40'	8	+11	20E
Pennsylvania					
Philadelphia	40°00'	75°10'	5	0	11W
Pittsburgh	40°26'	80°00'	5	+40	6.8W
Scranton	41°25'	75°40'	5	+3	11.4W
Rhode Island					
Providence	41°50'	71°25'	5	−14	15W
South Carolina					
Charleston	32°48'	79°58'	5	+20	4.5E
Columbia	34°00'	81°00'	5	+24	4E
South Dakota					
Aberdeen	45°28'	98°30'	6	+34	8.2E
Pierre	44°23'	100°20'	6	+41	9.7E
Rapid City	44°06'	103°14'	7	−7	11.8E
Tennessee					
Chattanooga	35°02'	85°18'	5	+41	1W
Johnson City	36°20'	82°23'	6	−31	3.5W
Knoxville	36°00'	83°57'	5	+36	2W
Memphis	35°10'	90°00'	6	0	3E
Texas					
Amarillo	35°14'	101°50'	6	+47	9.7E
Brownsville	25°54'	97°30'	6	+30	7.4E
Dallas/Forth Worth	32°47'	96°48'	6	+28	7E
El Paso	31°45'	106°30'	7	+6	11.2E
Houston	29°45'	95°25'	6	+12	6.4E
San Antonio	29°25'	98°30'	6	+34	7.7E
Utah					
Salt Lake City	40°45'	111°55'	7	+28	15.2E

	Latitude (North)	Longitude (West)	Time Zone	Add/Substract Minutes	Magnetic Variation (°) (East/West)
Vermont					
Bennington	42°54'	73°12'	5	−7	15.5W
Burlington	44°28'	73°14'	5	−7	15.5W
Montpelier	44°16'	72°34'	5	−10	15.5W
Virginia					
Norfolk	36°54'	76°18'	5	+5	8.7W
Richmond	37°34'	77°27'	5	+10	8W
Roanoke	37°15'	79°58'	5	+20	5.8W
Washington					
Olympia	47°03'	122°53'	8	+11	20.8E
Seattle	47°35'	122°20'	8	+9	21E
Spokane	47°40'	117°25'	8	−10	19.8E
West Virginia					
Charleston	38°23'	81°40'	5	+27	4.8W
Wisconsin					
Eau Claire	44°50'	91°30'	6	+5	2.3E
Madison	43°04'	89°22'	6	−3	1E
Milwaukee	43°03'	87°56'	6	−8	.5W
Wyoming					
Cheyenne	41°08'	104°50'	7	0	12.2E
Cody	44°31'	109°04'	7	+16	15.3E
Washington, D.C.	38°55'	77°00'	5	+8	9.2W
CANADA					
Edmonton	53°34'	113°25'	7	+32	20E
Halifax	44°38'	63°35'	4	+14	25W
Montreal	45°30'	73°36'	5	−6	16.4W
Regina	50°30'	104°38'	6	0	14E
Toronto	43°42'	79°25'	5	+18	9W
Vancouver	49°13'	123°06'	8	+12	22E
Winnipeg	49°53'	97°10'	6	+29	7.7E

MOTTOES

Add a little charm and personality to your dial with a motto.
This should be done before final finishing and setting.

Tempus fugit (Time flies)

Horam sole nolente nego (The hour I tell not, when the sun will not)

Life's but a walking shadow (from *Macbeth*)

Life's but a shadow
Man's but dust
The dyall sayes
Dy all we must

Now is between before and after

Now is yesterday's tomorrow

Sic transit hora (So passes the hour)

Tak tent o' time ere time be tint (Pay attention to time before it is wasted)

Time and tide tarry for no man

Ecce hora (Now or never)

Carpe diem (Seize the day)

Festina lente (Hasten slowly)

Dum licet utere (While time is given, use it)

Deus mihi lux (God is light to me)

Cosi la vita (Such is life)

C'est l'heure de bien faire (It is the time to do good deeds)

ADDITIONAL READING

By the time you reach this page, you already know a good deal about sundials, but if you wish to learn more—about the mathematics of dialing, the history of dials, and the place of sundials in man's conceptions of time—here is a list of books that will aid you in your quest to become a superior gnomonologist. All of these books are useful for different reasons, but let me make special note of two. The Mayalls' book is today a standard on the state of the art of sundial construction. It deals with simple graphics and geometric methods that require no mathematics, as well as with methods based on trigonometric functions. Albert Waugh's study, an accurate and eminently readable account of dialing, provides beginning and advanced dialists with much engaging information.

Popular

Earle, Alice Morse. *Sundials and Roses of Yesterday.* New York: Macmillan, 1902.

Gatty, Alfred (Mrs.). *The Book of Sun-Dials.* London: George Bell and Sons, 1889.

Mayall, R. N., and M. L. Mayall. *Sundials: How to Know, Use, and Make Them.* Cambridge, Mass.: Sky Publications, 1973.

Waugh, Albert E. *Sundials: Their Theory and Construction.* New York: Dover Publications, 1973.

Advanced

Cousins, Frank W. *Sundials.* London: John Baker, 1969.

Dolan, Winthrop W. *A Choice of Sundials.* Brattleboro, Vermont: The Stephen Greene Press, 1975.

Herbert, A. P. *Sundials Old and New.* London: Methuen, 1967.

Rohr, Rene R. J. *Sundials: History, Theory, and Practice.* Translated by Gabriel Godin. Toronto: University of Toronto Press, 1970.

Primitive Studies

Green, A. R. *Incised Dials or Mass Clocks.* London: Macmillan, 1926.

Horne, Dom Ethelbert. *Primitive Sundials or Scratch Dials.* Taunton, England, 1917.

Mottoes

Henslow, T. G. *Ye Sundial Booke.* London: W & G Foyle, 1935.

Hogg, W., and L. Cross. *The Book of Old Sundials.* London: T. N. Foulis, 1914.

Writers of Long Ago

Blagrave, John. *The Art of Dyalling.* London, 1609. Reprint. New Jersey: Walter J. Johnson, 1968.

Fale, Thomas. *Horologiographia: The Art of Dialling.* London, 1626. Facsimile edition. Ann Arbor, Michigan: University Microfilms.

Leadbetter, Charles. *Mechanick Dialling: The New Art of Shadows.* London, 1756. Facsimile edition. Ann Arbor, Michigan: University Microfilms.

Leybourn, William. *Dyalling.* London, 1700. Facsimile edition. Ann Arbor, Michigan: University Microfilms.

Time

Wright, Lawrence. *The Story of Time.* New York: Horizon Press, 1969.

Priestley, J. B. *Man and Time.* Garden City, N.Y.: Doubleday, 1964.

Sun

Herdeg, Walter, ed. *The Sun in Art* (no. 105). Zurich: Graphics Press, 1962.